AGROFORESTRY IN

A F R I C A

—— A SURVEY OF ——
PROJECT EXPERIENCE

AGROFORESTRY IN A F R I C A

— A SURVEY OF — PROJECT EXPERIENCE

PAUL KERKHOF

Edited by GERALD FOLEY and GEOFFREY BARNARD

PANOS

SALLY O'LEARY

Published by Panos Publications Ltd
Angel House,
9 White Lion Street,   2 APRIL
London N1 PD
UK

071- 278 1111

First published 1990

British Library Cataloguing in Publication Data:
Kerkhof, Paul
Agroforestry in Africa: a survey of project experience
1. Developing countries. Forestry
I. Title II.Barnard, G.W. (Geoffrey William) *1955-*
III. Foley, Gerald
634.9091724

ISBN 1-870670-16-7

The Panos Institute is an international information and policy studies institute,
dedicated to working in partnership with others towards greater public understanding
of sustainable development. Panos has offices in London, Paris and Washington
DC, and was founded in 1986 by the staff of Earthscan, which had undertaken
similar work since 1975.

For more information about Panos contact: Juliet Heller, Press Officer

Editorial director: Benjamin Pogrund
Production: Jacqueline Walkden
Publications director: Liz Carlile
Photo research: Patricia Lee
Graphic design: Viridian, London
Maps: Philip Davies
Illustrations: Carmen Miranda

Printed in Great Britain by Jolly and Barber Ltd, Rugby

FOREWORD

A great deal has been written about the potential for agroforestry. Examples of traditional agroforestry systems provide impressive demonstrations of how trees can be used on farms in highly productive and beneficial ways. Research trials with new agroforestry techniques and species combinations have also yielded encouraging results. On the strength of this, and amid high hopes, agroforestry projects have been started in many countries.

Living up to these expectations in practice is bound to be difficult. Agroforestry is still a very new discipline. To a large extent, project staff are having to develop the necessary techniques and approaches as they go. Often they are working in isolation. Although valuable lessons are being learned by projects every day, there are few mechanisms by which these can be shared with projects in other countries, or even in neighbouring districts.

This study is designed to help address this problem. It presents a survey of the experience of 21 projects in 11 countries throughout Africa. Although it makes use of project reports and other literature, it is based primarily on visits to each of the projects concerned. It looks at agroforestry from a pragmatic viewpoint; that of the people most directly involved in the design and implementation of projects. And it is to this audience that the study is directed.

A broad definition of agroforestry has been used and the projects have been chosen to represent a wide range of approaches and ecological conditions. What the projects have in common is that all contain elements which involve the active management of trees within the farming environment. Each has also been in existence for a minimum of 3-4 years — the average project age is nearly 9 years — so that there is enough experience for at least preliminary conclusions to be drawn.

The fieldwork was conducted between March 1988 and May 1989. Each project visit involved detailed discussions with project staff, local officials and others. The project profiles were then prepared and sent back to project staff for comments and corrections.

This final report is divided into three parts. **Part I** gives a summary of the overall lessons that have emerged from the experience to date. Profiles of individual projects are presented in **Part II**; the 21 projects visited are covered in 19 profiles, since two of them include two projects. To help make comparisons, profiles are grouped in five categories. Although there is some overlap between them, the projects in each category have a number of broad similarities in the environment in which they are working and in their approach and objectives. **Part III** draws upon the profiles to illustrate and discuss a number of key elements in the design and running of projects.

Funding for the study was provided by the Swedish International Development Authority (SIDA), the Technical Centre for Agricultural and Rural Cooperation (CTA), and the Commission of the European Communities. The main research and writing was carried out by Paul Kerkhof, an agroforestry specialist with working experience in several projects in Africa. The study was coordinated and edited by Gerald Foley and Geoffrey Barnard, of the Panos Institute.

A great many people are to be thanked for their assistance in preparing this study (see list of acknowledgements). Particular thanks are due to the staff of the projects visited. Without their patient explanations and the frankness they showed in discussing their work, this report would not have been possible. The advice and suggestions of those who reviewed drafts of the report are also gratefully acknowledged. Responsibility for the facts and opinions expressed here, however, rest solely with the authors.

ABBREVIATIONS

BAT	British American Tobacco Company
CARE	Cooperative for American Relief Everywhere
CTA	Centre Technique de Coopération Agricole et Rurale
CTFT	Centre Technique Forestier Tropicale
DANIDA	Danish International Development Agency
DGIS	Directorate General for International Cooperation (Netherlands)
ENDA	Environment and Development in the Third World (Zimbabwe)
EPAP	East Pokot Agricultural Project (Kenya)
FAO	Food and Agriculture Organization of the United Nations
FLUP	Forest Land Use Project (Niger)
GLUP	Gursum Land Use Project (Ethiopia)
GRAAP	Groupement de recherche et d'appui pour l'auto-promotion paysanne (Burkina Faso)
GTZ	Deutsche Gesellschaft für Technische Zusammenarbeit (German Agency for Technical Cooperation)
HADO	Hifadhi Ardhi Dodoma (soil conservation in Dodoma - Tanzania)
ICRAF	International Council for Research in Agroforestry
IDRC	International Development Research Centre (Canada)
IITA	International Institute for Tropical Agriculture
ILO	International Labour Office
INADES	Institut Africain de Developpement Economique et Sociale
KEA	Kondoa Eroded Area (Tanzania)
KWDP	Kenya Woodfuel Development Programme
NGO	non-governmental organisation
NORAD	Norwegian Agency for International Development
PAF	Projet Agro-forestier (Burkina Faso)
PAFSAT	Promotion of Adapted Farming Systems Based on Animal Traction (Cameroon)
PAP	Projet Agro Pastoral (Rwanda)
RAP	Rural Afforestation Project (Zimbabwe)
SECAP	Soil Erosion Control and Agroforestry Project (Tanzania)
SIDA	Swedish International Development Authority
SPU	Seedling Production Unit (KWDP)
TRDP	Turkana Rural Development Programme
UNHCR	United Nations High Commission for Refugees
USAID	United States Agency for International Development
ZOPP	Zielorientierte projecktplanung (target oriented project planning/planification participative par objectifs)

UNITS

1 kilometre = 0.62 miles
1 metre = 3.32 feet
1 hectare = 2.47 acres
1 square kilometre = 0.39 square miles

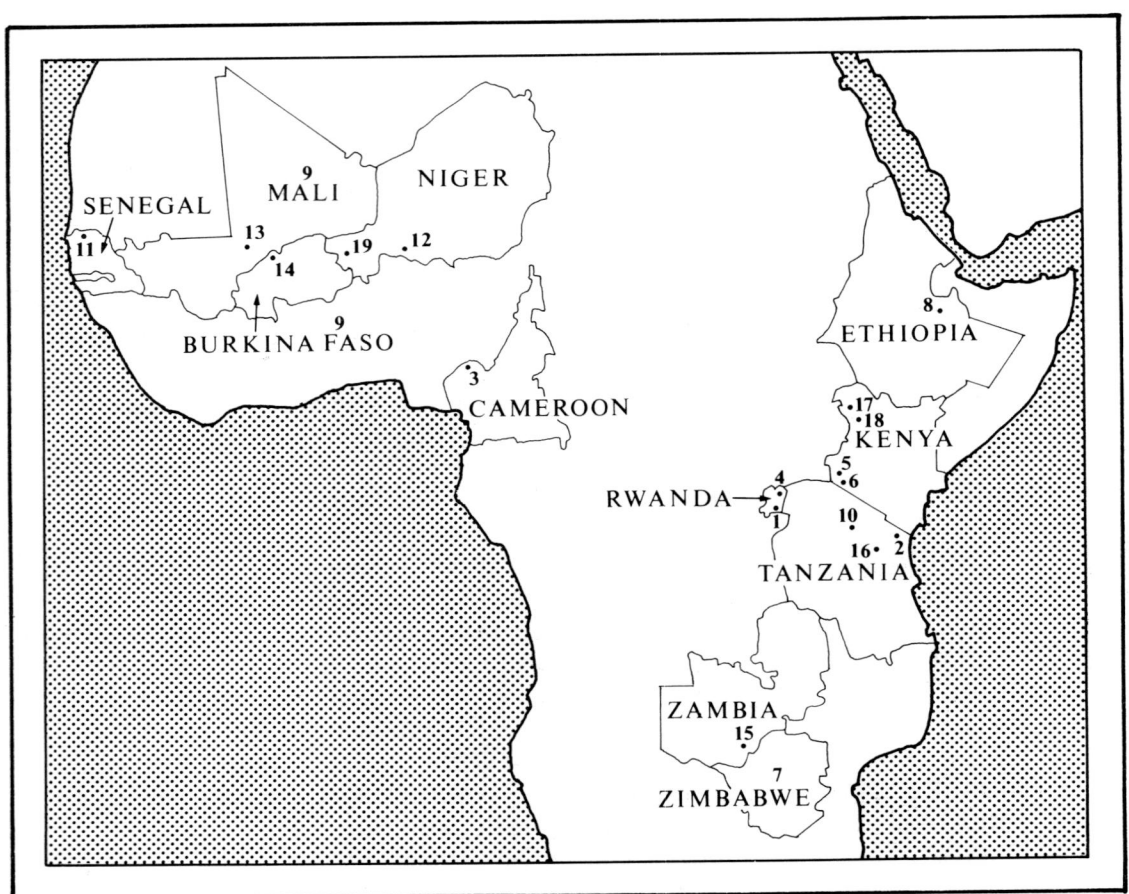

SENEGAL

MALI
9

NIGER

13

11

19 .12

14

BURKINA FASO
9

CAMEROON
3

ETHIOPIA
8.

.17
.18
KENYA

5
6

RWANDA
4
1

10

16 . 2

TANZANIA

ZAMBIA
15

ZIMBABWE
7

LIST OF PROJECTS COVERED IN SURVEY

Type A: Tree growing to increase productivity in high potential areas

Project	Short Name*	Official Name
1.	PAP	Projet Agropastoral de Nyabisindu,Rwanda
2.	SECAP	Soil Erosion Control & Agroforestry Project, Tanzania
3.	PAFSAT	Promotion of Adapted Farming Systems Based on Animal Traction, Cameroon

Type B: Tree growing for fuelwood and other products

4.	Gituza	Gituza Forestry Project, Rwanda
5.	KWDP	Kenya Woodfuel Development Programme, Kenya
6.	BAT	BAT Afforestation Project, Kenya
7.	RAP	Rural Afforestation Project, Zimbabwe
8.	Gursum	Gursum Land Use Project, Ethiopia

Type C: Village forestry projects

9.	Bois de Villages	Projet Bois de Village, Mali & Burkina Faso
10.	Village Afforestation	Village Afforestation Programme, Tanzania
11.	Ferlo	Reforestation Around Wells in Northern Senegal

Type D: Tree growing to increase productivity in dryland areas

12.	Majjia Valley	Majjia Valley Windbreak Project, Niger
13.	Koro	Koro Village Agroforestry Project, Mali
14.	PAF	Projet Agro-forestier, Burkina Faso
15.	Lusume	Soil Conservation and Agroforestry Project, Zambia

Type E: Projects to promote natural regeneration

16.	HADO	Hifadhi Ardhi Dodoma Project, Tanzania
17.	Turkana	Turkana Rural Development Project, Kenya
18.	EPAP	East Pokot Agricultural Project, Kenya
19.	Guesselbodi	Forest Land User Project, Niger

*To avoid repeating long official names, these names or acronyms are used in the text to identify projects

ACKNOWLEDGEMENTS

The authors would like to express their gratitude to the many people who helped in the preparation of this study. Special thanks are due to the following project staff, and to their colleagues, for the hospitality and cooperation the showed:

BURKINA FASO: Mr Daudi, Alice Didi, Sana Issaka, Matieu Ouadraogo, Wim Magerman, Jean Marc Perrier, Damas Poda.

CAMEROON: Karl Schleich, Mjei Mbah Wilfred, Klaus Zweier.

ETHIOPIA: Richard Doenges, Kevin Kamp.

KENYA: Floice Adoyo, Edmund Barrow, Noel Chavangi, Musa Enyola, Oscar Enyola, Mary Kekovole, Francis Kima, Simon Kimwe, Henry Kola, Sean McGovern, Duncan Michira, Jacob Moenga, Julius Mogire, Sarah Momenyi, Patrick Mungala, Milcah Ongayo, Boaz Shuma, Kees Vogt, Joseph Wekundah, Catherine Wituka, and to the extension workers and tobacco farmers of Kehancha.

MALI: Mr Kombogourou, Mr Koroporo, Dicko Adamas Mahamane, Donald Mansius, Michael O'Brien, Garibou Podiogo,

NIGER: Marshall Burke, Mani Djika, Armin Elbst, Leigh Heart, Amoul Kinni.

RWANDA: Ullrike Breitschuhe, Irma Corten, Peter Keller, Elie Hakizumwani, Mr Mashavu, Jurgen Schmitt-Duchard, Siegfried Schroder.

SENEGAL: Mr Badji, Ulf Baum, Siggy Tluczykont.

TANZANIA: John Baributsa, Carl Gerden, Mr Haule, Lars Johansson, Mwito Matiko, Mr Mbegu, Michael Mdoe, Mr K.C.H. Mndeme, Frank Mongi, Mr Njau, Mr Nkwera, Kiki Norden-Olsson, Cassian Sianga, Mr Sakaya.

ZAMBIA: Yembo Kaonga, Frank Boehner.

ZIMBABWE: Blake Chakavanda, Antony Chazambamwe, Mr Chihambakwe, Jeanette Clarke, Davison Gumbo, Yemi Katerere, Mr Phiri.

We would also like to thank the following people for their helpful advice during the project, and for their comments on earlier drafts:
Gordon Armstrong, Mike Arnold, Michel Baumer, Kjell Christophersen, Charles Condamines, Jean Clement, Alistair Fraser, Peggy Fry, Olivia Grant, Karin Holmgren, Dominic Hounkonnou, Robert Huggan, Nils Kjølsen, John Michael Kramer, Jan Kuijper, Philippe de Leener, Mike McCall, Mathenge Mnene, Phil O'Keefe, Gay Pedlaw, John Raintree, Chris Rey, Marion van Schaik, El Hadji Sene, Gill Shepherd, Margaret McCall Skutsch, Claire Vignon, Remko Vonk, Karin Wohlin.

CONTENTS

PART I

LESSONS FROM EXPERIENCE

LESSONS FROM EXPERIENCE

The projects covered in this study represent a broad spectrum of agroforestry experience. In selecting and analysing them, the emphasis has been on the practical results achieved. The aim has been to find out what has worked, and what has not worked, under actual project conditions.

Taken together, the 21 projects have been running for a total of nearly 180 years. Although experiences have varied widely, a number of clear lessons are emerging. Some are extremely positive. But it is also important to point out what has failed or not gone well. These provide a warning to project designers; they are also a challenge to those working on research and development. The future of agroforestry depends on developing techniques and approaches which are robust and well-proven, and which can be applied under the real-life conditions faced by field projects.

ATTITUDES TO TREE GROWING

Whatever the climatic and ecological conditions, most people are well aware of the benefits of trees. What varies is their attitude to growing them.

In areas where the soils are fertile and the climate is favourable, most projects have found that farmers are already growing a substantial number of trees. This is true even in areas where population densities are extremely high and the size of farms is small. Projects which have provided farmers with seedlings of the species they want, or in some cases just seeds, have generally been able to bring about significant increases in tree growing.

New species and cultivation techniques have also been well received by many farmers in these areas. There is, however, a widespread resistance to radical changes in farming techniques, especially where these require a high input of labour. For obvious and entirely rational reasons, farmers tend to prefer small, gradual changes in their farming methods. And unless they are convinced of new techniques, they are unwilling to take any major risks in trying them out.

In dryland farming areas, growing trees is much more difficult. Growth rates are slower, survival rates are poorer, and protection of seedlings is more of a problem. Farmers are therefore more cautious about investing time and effort in tree growing. While projects have found that farmers are willing to plant some trees, the uptake of seedlings from nurseries has often been disappointing. Some projects are finding that encouraging natural regeneration is more acceptable to farmers than tree planting because it is cheaper and less risky. Some have also broadened their focus and are now promoting contour bunds and other physical techniques aimed at improving agricultural production.

In pastoral areas, tree planting efforts have had poor results almost everywhere. The costs involved are high, and growth rates are usually very low. Protecting areas from grazing animals has been found to be a much more effective way of restoring the natural vegetation and tree cover.

The market for construction wood provides a strong motivation to grow trees; in Tanzania, **Grevillea robusta** *is widely planted for this purpose.*

The key to such schemes is control over grazing rights. This is a sensitive subject in any pastoral community. Some projects have managed to introduce controls through close collaboration with local communities; others have imposed them with the help of rigid government regulations and strict policing. Where controls have been effective, the impact on the local vegetation has often been dramatic, with large areas of severely degraded land being brought back into productive use at a fraction of the cost of any equivalent tree planting scheme.

INCENTIVES

A number of projects have used food or cash payments to encourage people to plant trees or carry out soil conservation and other work. Although it is questionable whether farmers will be motivated enough to continue this work once the incentives are withdrawn, the physical achievements of these projects have often been considerable.

Where farmers have chosen voluntarily to grow trees on their land they have generally been very selective in what species they have planted. Their views on species choice have often varied widely from those of project staff; nitrogen fixing species, for example, have often been less popular than projects initially assumed. Projects which have responded to such preferences have generally had a much better response than those which have retained fixed ideas about what trees should be grown.

Project experience also shows clearly that fuelwood scarcities, by themselves, rarely provide a sufficient incentive for people to plant trees. Although fuelwood may be valued as a secondary product, farmers are much more interested in trees as a source of construction wood, poles, fruit, and other products — especially when these can be sold for cash; other uses such as shade and the marking of farm boundaries are also important for some farmers. Projects which have had a narrow fuelwood focus have therefore faced problems.

EFFECT OF TREES ON CROP PRODUCTION

The use of agroforestry techniques to boost crop production has been the explicit aim of several projects. Up to now, however, none have been able to provide hard proof that this can be achieved under field conditions.

Promising results have been obtained from intercropping trials in the PAP project in Rwanda but these have not yet been confirmed in farmers' fields. Similarly, there is some evidence that the windbreaks in the Majjia Valley project in Niger have led to a net increase in crop production. But the results are very variable and, up to now, it has not been possible to draw any firm conclusions.

The fact is that trees take up land which might be used for crop growing. They also compete with adjacent crops for light and soil nutrients. Any increase in crop yields in the overall system must be sufficient to outweigh these negative effects if the farmer is to obtain a net benefit.

The general message emerging is therefore a cautious one. While positive effects on crop yields may be occurring in some cases, they certainly cannot be guaranteed. Unless there is specific evidence to go on, it is dangerous to make any assumptions about improved crop yields. It is better

Firewood shortages rarely provide an adequate incentive to plant trees; but when trees are grown, firewood is seen as a useful by-product.

to view projects in a broader context which takes into account their potential for providing tree products, environmental improvements and other benefits.

5

THE NEED FOR FLEXIBILITY

A criticism that can be applied to many of the projects is that they began with preconceived ideas about what the local problems were and how they should be solved. In practice, these assumptions have often proved wrong. To their credit, most projects have recognised this sooner or later, and changed their approach; in some cases it has meant launching an almost completely new project.

This flexibility has been crucial to the success that projects have achieved. There is no doubt, however, that a great deal of effort and resources have been wasted in pursuing approaches which were not well adapted to local realities. Pressure to produce quick results is often the main reason for this. But, as experience clearly shows, the risks in pushing ahead with projects before the initial assumptions have been properly tested are bound to be high.

Experiences in the Majjia Valley in Niger suggests that windbreaks can increase crop yields, but results vary a lot from year to year, and from one farm to another.

DESIGNING SURVEYS AND RESEARCH TRIALS

Well-designed initial surveys, followed up by proper monitoring and evaluation, are crucial in putting projects on the right track, and ensuring they stay there. In practice, many of the projects visited did not carry out surveys until well after they had begun, and few have succeeded in establishing effective monitoring and evaluation mechanisms.

Getting the right balance when designing surveys, either for collecting baseline data or for subsequent monitoring and evaluation purposes, is not easy. Where surveys have been carried out, there has been a tendency for the techniques involved to become over-elaborate, requiring lengthy questionnaires and a large amount of data analysis. At the same time, important issues such as land tenure, tree ownership rights, labour availability, and the role of women in decision-making, have often been ignored.

While detailed surveys are clearly useful in some circumstances, there is a danger in trying to do too much. Surveys can soak up an enormous amount of time and effort, and yield few practical results. Where resources are limited, simplicity has generally proved to be the best approach. But for this to work, surveys need to be carefully designed so that they focus as closely as possible on the issues most directly relevant to the project.

It is also important to ensure that surveys are treated as an integral part of the project and not as a separate academic exercise. Involving extension staff in carrying out surveys, and in analysing their results, is usually the best way of doing this; it also has a useful staff training function.

Similar conclusions apply to research trials. They need to be integrated into the project if they are to provide usable results. They also need to be matched to the resources, timescale and requirements of the project. There is little point in beginning elaborate trials that cannot be followed up because of lack of funds or manpower, or which will not provide results in time to influence project strategies.

The question of the relative merits of "on-station" and "on-farm" research is still being hotly debated in agroforestry circles. It is extremely difficult to design trials so that they provide statistically valid results, while, at the same time, accurately reflecting the complexity and variability of local farming conditions. There is no easy way around this problem, so a combination of both on-farm and on-station trials will often be needed.

PROVING THE TECHNICAL PACKAGE

One of the clearest lessons from the projects is the need to test the technical package under local conditions before starting to disseminate it. Attempts to transfer agroforestry techniques directly from one project to another, or from the text book or research station to the field, have often been disappointing.

Geoff Barnard/Panos Pictures

Among the projects visited, the most obvious example encountered was alley cropping with *Leucaena leucocephala*. This has been the most widely publicised agroforestry package and has, at times, been treated as though it were a universally applicable technique. In fact, in its original form, it was a major disappointment in all the projects in which it was tried.

This is not to say that agroforestry techniques such as alley cropping cannot ever be transferred from one place to another. It simply emphasises the point that packages must be shown to work under local conditions before they are used as the basis of a large-scale extension effort.

In many cases, technical packages need to be modified to suit local conditions. With alley cropping, for example, some projects have found that farmers are prepared to take the basic idea and adapt it to their own circumstances. They have used different species and instead of planting closely-spaced rows of trees they have turned it into a form of dispersed intercropping — a practice which has generally proved to be much more

Alley cropping has shown impressive results in research trials; but among the projects visited, farmers preferred more dispersed forms of intercropping.

7

popular with farmers than more intensive forms of intercropping.

Where projects do not have the time or resources to mount a major research programme, enlisting the help of farmers as "local researchers" can be an effective alternative. They can be offered a range of techniques and species and it can be left to them to find out which they find most useful. Once this screening has been done, the most popular packages can then be used with confidence as the basis for more widespread extension efforts.

EXTENSION METHODS

All projects have had to face the difficult task of communicating their message to local people. A wide range of extension approaches have been used. None have been wholly effective, and most projects recognise that a lot more needs to be done in some areas, for example, in drawing women into projects.

The most important point is that extension staff must get out into the field and meet people if they are to do their job properly. This is often difficult because roads are bad, transport is scarce, and field conditions are not easy. But it marks the fundamental difference between successful and unsuccessful extension work.

Headquarters staff must support this process by getting involved themselves, meeting field staff and providing them with the necessary support and encouragement. It is obviously important to have procedures for project planning and reporting. But if these end up disrupting the basic job of meeting farmers they can turn out to be counter-productive.

It is also essential that extension approaches are matched to local customs and traditions. Some projects have used village meetings to spread the project message, while others have worked through women's groups, schools and churches. A number of projects have worked with individual farmers; the disadvantage of this approach is that it is slower and less cost-effective than group extension.

Many projects have found that audio-visual techniques such as the use of slide shows, pictures and drama, have been helpful in attracting local interest. But it is important to realise that simply entertaining people is not enough; if they are to be influenced by the show, the examples chosen must be relevant to local people and carefully tuned to their own problems and realities.

Transferring technical packages from one place to another is not easy; **Leucaena** *grows well on this farm in Tanzania, but elsewhere its performance has often been disappointing.*

8

Chris Pennarts

Schools have been used by many projects to spread the project message.

Model farms have been used by a number of projects to demonstrate new techniques, but in most cases they have had very little impact on local farmers. Although they may have worked well at a technical level, and looked impressive to visitors, model farms have remained isolated islands in the farming community. As a result, most projects have now abandoned this approach.

In the majority of projects, emphasis has been placed upon the need for local participation. A number of projects have evolved from a heavily centralised attitude towards one in which local people are increasingly encouraged to define how they wish the project to assist them. It is worth noting, however, that this is not the only approach which has worked. Several projects have used more authoritarian measures, such as enforced grazing controls, with considerable success.

INSTITUTIONAL FRAMEWORK

The institutional background of projects has varied. In terms of their success rate, no differences were found between projects managed by government extension agencies and those coordinated by NGOs, either locally-based or international. Some of the best results have emerged from projects where there has been effective collaboration between NGOs and government agencies.

Neither is there any correlation between the success of projects and the amount of money spent on them. If anything, the most expensive programmes, with budgets of US$500,000 per year and more, seem to have a poorer record than those costing less. This may be because the larger projects have tended to invest more heavily in offices, staff housing, vehicles and other items with a corresponding lack of emphasis on the practical aspects of achieving results in the field.

An important question raised by several projects is whether forest departments are the best agencies to deal with agroforestry extension. In countries where foresters are still seen as policemen, this presents a major barrier to winning the trust of local people. Some forestry departments also lack the training and infrastructure required to mount a major agroforestry extension programme.

In such cases, there are arguments for saying that agricultural extension agencies may be better placed to promote agroforestry. Certainly, there is a strong case almost everywhere for broadening project teams to include agriculturalists and other disciplines, instead of relying entirely on foresters.

Sustainability is a key question for all projects; who will look after nurseries, for example, once project funds dry up?

Whatever agency is implementing a project, it is important that support is guaranteed over a sufficiently long period. Agroforestry projects, by their very nature, take time to develop and are bound to be slow to produce results. A commitment of at least eight years is essential if projects are to be given a realistic chance of success.

It is significant that most of the projects visited had to change direction, in some cases quite radically, as a result of their first few years' experience. This underlines how important it is that project staff are allowed to learn from experience without being under constant pressure to produce immediate results.

CONCLUSIONS

No final judgment is yet possible on any of the projects visited in this study. Most have been successful to some degree, although not always in ways that were originally anticipated. At the same time, all of the projects still face a variety of problems and challenges and recognise that a great deal of work remains to be done.

On the key question of sustainability, few projects have yet reached the stage where they can confidently predict that the changes they have introduced will continue, and spread, after the project has finished. For some, especially those involving a large food-for-work element, this may be an unrealistic goal. But there is no question that, although mistakes have been made, many of the projects surveyed have made substantial progress and are heading in a useful and positive direction.

The study demonstrates, therefore, both the potential for agroforestry, and the difficulties involved in putting it into practice. It emphasises that there are no simple agroforestry solutions that can be expected to have instant results, or be universally applicable. But it reinforces the basic relevance of agroforestry as an approach, provided it is viewed in a broad sense, and provided it is implemented in a way that is sensitive to the needs and priorities of local people.

PART II

THE PROJECTS

PROJET AGROPASTORAL DE NYABISINDU, Rwanda

FROM ECO-FARMS TO AGROFORESTRY

The need to introduce intensified farming techniques to Rwanda has long been evident. The population densities in parts of the country are among the highest in Africa and are stressing the existing agricultural system near to breaking point.

Attempts have been made to introduce a variety of improved agricultural methods to the project area over the past 20 years. Leguminous shrubs are given a high priority by the project for ecological reasons, but planting *Grevillea robusta* for timber has proved much more popular with the local farmers.

Promoting eco-development

The Nyabisindu area is in southern Rwanda. Like most of the rest of the country, the terrain is hilly with an altitude varying between 1,500 and 2,000 metres above sea level. The annual rainfall is about 1,200mm.

Since the area is close to the mountain rainforest zone, the vegetation is lush and green and gives the impression of a flourishing farming area. But Rwanda is one of the poorest and most densely populated countries on earth. The six communes presently covered by the project have a total population of 300,000 people; the communes are divided in turn into sectors, each with a population of about 5,000 people. The population density exceeds 400 persons per square kilometre in parts of the project area and is still growing.

PAP has been running since 1969. It is thus one of the oldest and best known agroforestry projects in Africa. An average of about 2,000 people visit it each year.

Since the beginning, it has been supported by GTZ. The Government of Rwanda contributes about 20% of the total budget, excluding staff salaries. GTZ also provides 3-4 permanent technical advisors.

The project began with the provision of funds for the rehabilitation of a milk

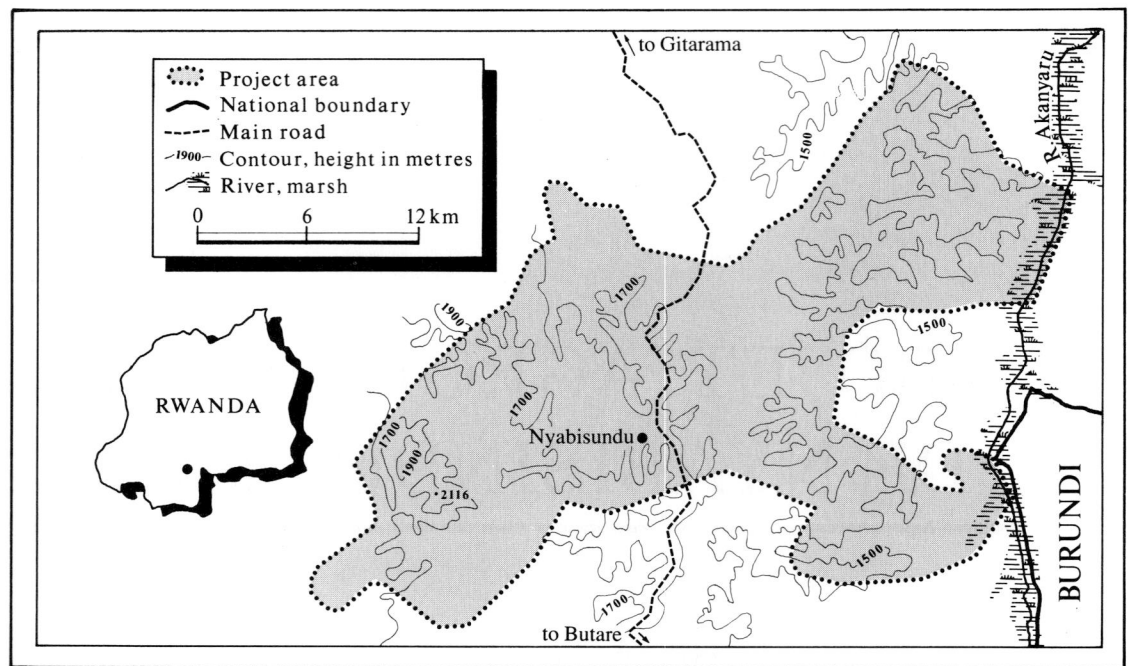

Name of project:	**Projet Agro Pastoral de Nyabisindu (PAP)**
Address:	**BP 70, Nyabisindu, Rwanda**
Project area:	**6 communes in southern Rwanda**
Average rainfall:	**1,200 mm per year**
Implementation:	**Ministry of Agriculture with 3-4 expatriate advisors**
Funding: up to 1987:	**US$ 760,000 per year**
	in 1988: US$ 540,000
	All external funding provided by GTZ
Exchange Rate:	**80 Rwanda Francs = 1 US$ (October 89)**

factory and the improvement of local milk collection. After the oil crisis of 1973, project staff observed that development in the area was inhibited by problems such as high fertiliser prices and fuelwood shortages. The project was then extended to cover fodder production, subsistence food cultivation and tree growing.

The project philosophy was based on the ecological ideas developed during the 1970s. Reacting against the green revolution approaches popular at the time, it argued in favour of organic farming as a means of restoring and maintaining the declining soil fertility in the area. In the system of "eco-development" promoted by the project, all the components were intended to be linked together in a mutually supportive and self-sustaining manner.

Some 80-100 model farms based upon these organic farming principles were established. Paid extension workers were used to explain the new techniques and farmers were provided with all the materials and, if necessary, the casual labour required to create the farms. This level of intervention was regarded as essential if the project ideas, which required a simultaneous transformation of the whole farming approach, were to be realised.

It was believed that the model farms, if they succeeded, would automatically act as centres of dissemination. The project staff argued at the time that, if the proposed technical packages were good enough, neighbouring farmers would spontaneously start using them.

The strategy of creating and supporting model farms was followed until the late 1970s, when evaluations revealed that little diffusion was taking place. The farms themselves had obvious merits from the project viewpoint. They were relatively easy to control, they were convincing to visitors, and they were reassuring to funders. But whatever their merits, most remained isolated islands in the farming community.

This phase of the project did, however, have an important effect on the general awareness of ecological constraints on rural development in Africa and elsewhere. It highlighted the limits of the green revolution and it helped in the development of alternatives, such as agroforestry, green manuring and zero grazing.

A major agroforestry research programme

In 1979, a decision was made to set up a research programme to provide a firm scientific basis for the project's recommendations. A team of three expatriate agroforestry research advisors was employed and they initiated one of the most extensive sets of tree and foodcrop trials in Africa to date.

These have produced a wealth of useful data (Neumann and Pietrowicz, 1986). The most successful agroforestry system designed by the project uses rows of *Grevillea robusta* planted in "soil conservation strips" on the contours of sloping ground and along farm boundaries. In the project model they are usually associated with an under-storey of leguminous trees and grasses. The direction of the strips is preferably east-west, to reduce shading of foodcrops, unless the orientation of the contours requires otherwise.

The project uses hand-painted pictures to illustrate new techniques.

Various densities and rotation periods were tried. The highest combined production from the trees and food crops was obtained with 400-600 *Grevillea* per hectare at an age of 4-6 years, or 250-300 trees per hectare at 9-10 years. In both cases the maximum crown cover is about 20%.

In one trial, with 350 *Grevillea* per hectare, the annual yield after

15

nine years was 14.6 cubic metres of wood and 3.07 tonnes of leaves (fresh weight). The researchers argued that such yields were more than sufficient to meet the annual wood needs of a family of six. They added, however, that rigorous pruning and thinning are required if food yields are to be maintained with such an intensive intercropping system.

Drawing showing how tree prunings can be used for mulch.

Numerous trials were also carried out with dispersed intercropping of leguminous species. The results varied greatly depending on the species, site and other factors. In the case of *Sesbania sesban*, they showed that, when intercropped with coffee at a density of 500 trees per hectare, total biomass production from the trees amounted to 7.4 tonnes per hectare (fresh weight) after nearly 4 years. A major advantage of this species is that they do not require strict management. Unlike *Grevillea*, late pruning or harvesting of *Sesbania sesban* has relatively little impact on food crop yields.

The trials also found that trees in pasture had a much lower survival rate and grew more slowly than those in croplands. This was especially so on marginal soils where four times as many trees survive in cropland as in pasture. The growth in the height of trees in cropland was also 1.5-3.0 times faster than in pasture lands. This is perhaps not surprising given the root competition in grassland, while cropland is kept weed free. Similar results have been found by many other projects elsewhere in Africa.

Table 1.1 The effect of pruning height on yield of leguminous hedges.

Species	Pruning height (cm)	Production* (kg)
Leucaena leucocephala	80	102
	45	66
Calliandra calothyrsus	165	196
	50	124

* *Production figures measured in kilograms of total dry matter per 100 metres of hedge per year.*

Source: Neumann and Pietrowicz (1986)

16

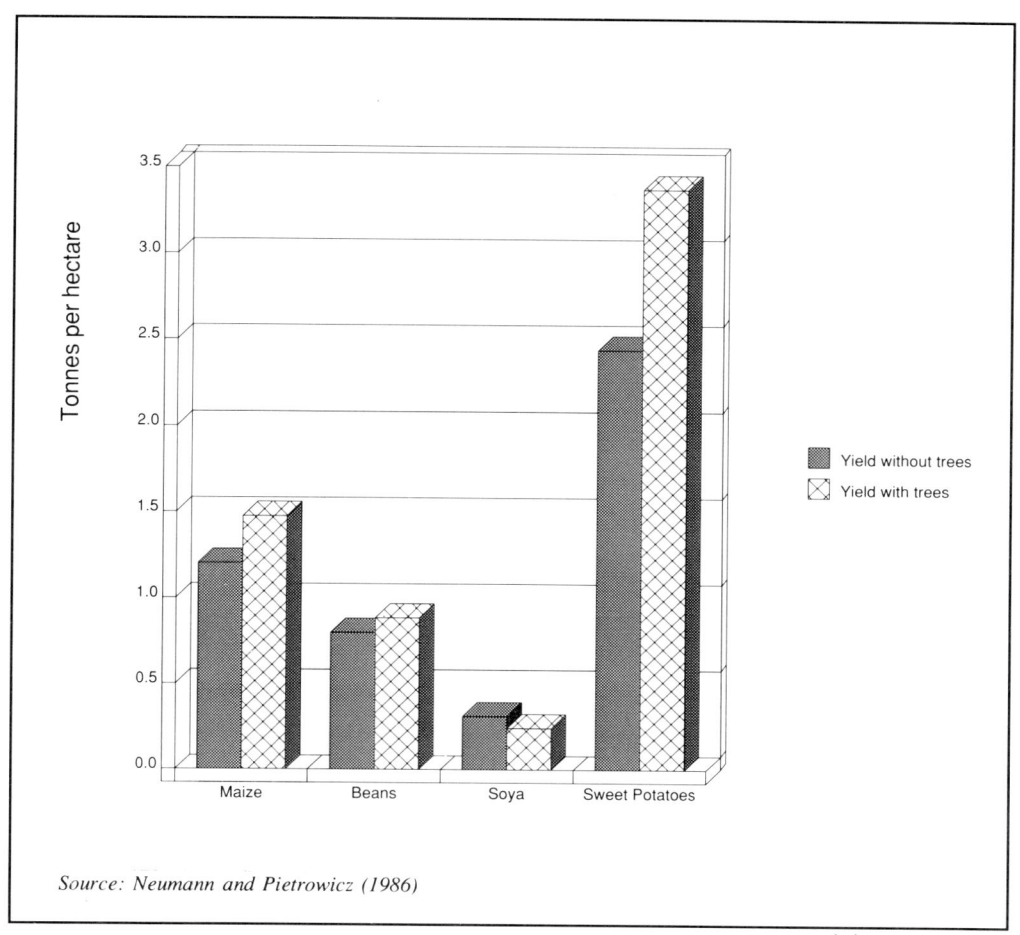

Figure 1.1 Effect of intercropping on crop yields (tonnnes per hectare, average over five growing seasons).

Trials were also carried out with hedges. The species used included *Calliandra calothyrsus, Leucaena leucocephala, Sesbania sesban, Morus alba* and *Tephrosia vogelii*. The results suggested that the hedges should be planted as 60cm wide strips on the high side of terraces and contour bunds. They also indicated that a pruning height of a metre or more was considerably better than cutting just above the ground. Table 1.1 shows some of the annual dry matter production figures for different species and pruning heights.

The effects of intercropping with trees on the yield of food crops were measured over a number of years, and were generally found to be positive. Some of the results obtained are shown in Figure 1.1. These show increased yields with maize, beans and sweet potatoes, but a decrease in the case of soya.

This intensive research programme led to the formulation of a number of project packages. *Grevillea robusta* is recommended for intercropping as well as for planting along field boundaries and contours; the density of the trees should be reduced with successive rotations. Leguminous shrubs are primarily recommended for planting in 60cm wide hedges on the high side of contour lines and terraces.

Project researchers have calculated that the financial return from a farm using these techniques would be almost twice that from a farm without them. Whether farmers find this convincing, or are able to replicate it in practice, remains to be seen.

During the period 1980-1984, the trials were carried out on-station or under controlled conditions on model farms. The researchers believed that it was virtually impossible to obtain scientifically reliable data from farmer-managed trials. Since then, the emphasis has been shifted to on-farm trials. Despite the thoroughness of the on-station research, PAP now believes that future research must be carried out under actual farming conditions.

Dissemination techniques

The use of model farms as a basis for extension was dropped in 1982. The project then adopted the extension method developed in the Projet Agropastoral de Kibuye in west Rwanda. The new approach became fully operational from 1985 onwards.

In the new system, the appropriateness of packages developed in the research phase is first tested on groups of farmers in a pre-extension phase. The total number of farmers involved varies from 200 to 1,000. Inappropriate techniques or extension methods can therefore be corrected before mass extension takes place.

Extension relies principally upon village meetings at which the extension agent presents a particular theme using colourful locally produced pictures on a flannel board. The aim is to start a discussion on the subject among the villagers. The themes are established on the basis of annual consultations and interviews with farmers. The discussion is followed up with the distribution of leaflets in the local language and, sometimes materials such as seeds or seedlings. Although this approach has its own character, it is similar to the GRAAP method developed from 1979 in Burkina Faso.

Among the themes presented are environmental degradation, contour strips, and organic fertilisers. A typical series contains about 15 pictures. Because of the large number of extension workers, thousands of coloured pictures have to be produced. The project runs a workshop in which up to 10 staff are employed in painting the pictures.

During the 1970s, PAP had its own extension workers but from 1980 it has built on the existing extension service of the Ministry of Agriculture. Extension activities have, however, been hampered by a lack of skilled personnel. Moreover, the available extension workers tend to be diverted from agricultural extension by the local authorities and used for other duties such as imposing fines and collecting taxes.

The total number of agricultural extension staff in the project area is about 70 and there are six forestry extension workers. Overall supervision is provided by agronomists, of whom there is one per commune. There is one forester for the project area. There are also some veterinary staff.

The extension workers receive about three weeks' training annually from the project, and get some additional agricultural training from INADES, an agricultural training institute for Francophone Africa. The forester, agronomists and veterinary staff receive 3-4 weeks.

Strengthening planning capabilities

The lack of adequate planning, co-ordination and control in the Ministry of Agriculture has been one of the major constraints on the extension activities of the project. A major effort has therefore been made to strengthen the Ministry.

As part of this, the project has introduced an elaborate planning system which endeavours to ensure that project objectives, the means used to reach them, and the actual outputs achieved are continuously assessed and verified. The system is called ZOPP (ziel orientierte projekt planung, or "target oriented project planning") and is based on that developed by GTZ for its own projects (GTZ, 1987).

Under the ZOPP system, project planning is based on discussions at all levels, from farmers to

Increased production of fodder for stall feeding is one of the project's main objectives.

project management. For the 1986 work programme, for instance, over 100 meetings were held involving farmers, extension workers, agronomists and project management. A needs assessment was carried out among 130 farmers, 440 farmers were approached in the preparation of training courses, and separate group meetings were held for women.

Each year, all the project objectives, targets, verifiable indicators of progress, actions to be taken, methods to be used, and personnel responsible are written down. The annual plan continues for some 60 pages in this extremely thorough manner. Extension workers and other staff make work-plans which are much simpler, but nevertheless they must be done on an annual, quarterly, monthly and daily basis. Performance is continuously assessed by means of the monitoring and evaluation system of the project. As a result, it is now soon discovered if a member of the extension staff is spending his time with farmers, or in a local bar.

The ZOPP method has been used for other projects and is being considered by the government for use throughout the country. The system, however, requires highly skilled personnel and presumes that computer processing facilities are available. Without the necessary staff, equipment, and a considerable degree of discipline, it would be unlikely to work in other regions of the country.

Replacement of eucalyptus plantations

Eucalyptus has been planted in Rwanda for a long time and is still a very popular tree species. The cropping of eucalyptus on a short rotation, however, especially if all the biomass is removed, leads to rapid depletion of the reserves of nutrients in the soil. As a result of this, together with over-grazing, many of the old eucalyptus stands have become unproductive. In some cases, the undergrowth has completely disapeared and there is severe soil erosion, with up to a half a metre of soil washed away.

This is not specifically because the stands consist of eucalyptus, as is sometimes thought. It is rather a result of the way the trees are harvested, the fact that they grow so quickly and the overgrazing which has been permitted. Much the same problems would occur with any other highly productive crop under similar conditions.

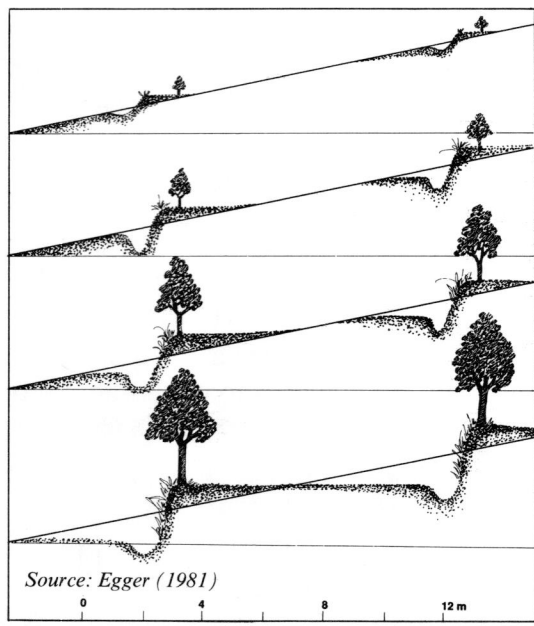

Source: Egger (1981)

Figure 1.2 Trees are planted on contour bunds; the idea is that over time natural terraces will form.

One of the project components is the conversion of these exhausted communal eucalyptus woodlots into improved plantations. The communal input is the cutting of the trees and establishment of anti-erosion structures on the contour lines. The project then removes the stumps and plants the area. Usually *Pinus* species are used to replace the eucalyptus. They are among the few tree species that will grow on the degraded soils, although they cannot be expected to restore the fertility of the soil.

Table 1.2 Project results in 1986	
Number of farmers trained:	42,397
Percentage of farms using soil conservation structures:	
Rukondo, Karama communes	80%
Kigoma, Murama, Nyabisindu communes	30%
Ntongwe, Ntyazo communes	15%
Number of farmers practising composting:	90%
Number of farmers receiving improved seeds	11,109
Number of nurseries	124
Total number of tree seedlings produced	2,388,190

Source: PAP (1987)

Table 1.3. Tree growing practices among farmers

Tree growing approach	Percentage of farmers using this approach	Average number of trees per farmer
Trees in cropland	72%	10-50
Trees on contour lines	80%	72
Trees in woodlots	73%	200

Source: Corsten (1989)

Decentralisation of seedling production

During its early stages, the project established and managed a large number of tree nurseries. These were tightly run by the expatriate advisers and thus did not provide a model for sustainable local management. The project, however, had little alternative if it wanted to produce seedlings; trained Rwandan staff were very scarce in the 1970s because virtually no education had been provided during the colonial era.

By 1986, a total of 124 nurseries had been established and were producing 2.4 million seedlings of various kinds. As a result, most farmers had relatively easy access to the kind of seedlings they wanted.

Managing such a large number of widely dispersed tree nurseries is, however, difficult and since 1988 the project has changed its seedling production system. Responsibility for the nurseries has been handed over to private operators and the project pays for the seedlings delivered. Under the contract between PAP and the nursery operator, a series of payments are made during the production cycle, such as when the seeds germinate, when they are pricked out and when they are finally delivered.

Monitoring and evaluation of the project impact

As a part of the ZOPP system, monitoring of project activities is regularly carried out. The number of training sessions, for example, the number of participants in the sessions, the quantity of information material distributed and the number of farmers who adopted a particular innovation are closely monitored. Small surveys are frequently held for this purpose. Some of the results from 1986 are shown in Table 1.2.

Although monitoring and evaluation of the project activities is generally thorough, there was, until recently, little follow-up of seedling survival and performance. A survey in 1988, however, showed that 70% of the trees were in good shape with 30% dead or in poor condition.

An overall survey of tree growing practices in the project area was carried out in 1987. This found that tree growing was extremely common among farmers, as shown in Table 1.3.

In an attempt to establish the extent to which such tree growing was a result of traditional practices or had been stimulated by the project, the survey asked farmers how much they had planted in the past compared with the present. This revealed that few farmers had intercropped before 1975, more did so after 1980 and a majority of farmers after 1983. Although such figures, based on memory recall, are notoriously unreliable, in this case they can be broadly substantiated by looking at the age of the trees. It thus seems clear that the project has had a substantial impact on tree growing in the area.

One of the most noticeable features of the area is, in fact, the vast number of timber trees, of which *Grevillea* is by far the most common. Most farmers have also planted a few leguminous shrubs, if only to relieve the continuing pressure from extension staff. Project staff accept that the benefits from the shrubs bear little comparison with those from *Grevillea*, a view evidently shared by the farmers. When seedlings are distributed, everything is taken except for the leguminous shrubs. After 20 years' experience of trying to promote their use, it is a point worth bearing in mind.

References:

CORTEN, I. (1989). "PAP onderzoeks gegevens".

GTZ (1987). "ZOPP Initiations aux éléments de la méthode".

NEUMANN, I. and P. PIETROWICZ (1986). "Agroforesterie à Nyabisindu - Etudes et experiences No. 9". PAP, Rwanda.

PAP (1987). "Realisations de l'exercise 1986". PAP, Rwanda.

SOIL EROSION CONTROL AND AGROFORESTRY PROJECT, Tanzania

TAKING AN INTEGRATED APPROACH

The mountainous West Usambaras area in Tanzania has suffered severely from deforestation and overgrazing in recent decades. A programme to combat these problems has been under way since 1981.

Efforts have been made to introduce anti-erosion measures and improved methods of dairy farming. A reforestation programme has been launched and forest management plans have been drawn up in collaboration with local communities. The preliminary results show that some of these measures are quite successful; others, however, do not make economic sense for the farmers.

The project area

The West Usambaras are a mountain range in the north-east of Tanzania. They vary in altitude from 1,400 metres above sea level in the valleys up to about 2,200 metres on the upper mountain slopes. The surrounding lowland plains are only a few hundred metres above sea level. Much of the area was formerly covered with dense forests but over the past decades these have been largely cleared.

The administrative headquarters of the district is the town of Lushoto, the first capital of the colony of Tanganyika. It has an old forest research station which is now part of the Tanzanian Forestry Research Institute. Rainfall measurements have been carried out here since the early part of the century and show a steady decline from about 1,300mm per year in 1910 to 1,100mm in recent years.

The mountains cover a total of 4,500 square kilometres, of which about 70% is under crop cultivation. The inhabitants are settled farmers from the Washambaa, Wapare, and Wambugu peoples. The total population is about 300,000 and is growing at about 3% per year. The average population density is 68 persons per square kilometre but it reaches 400 persons in some of the heavily cultivated areas.

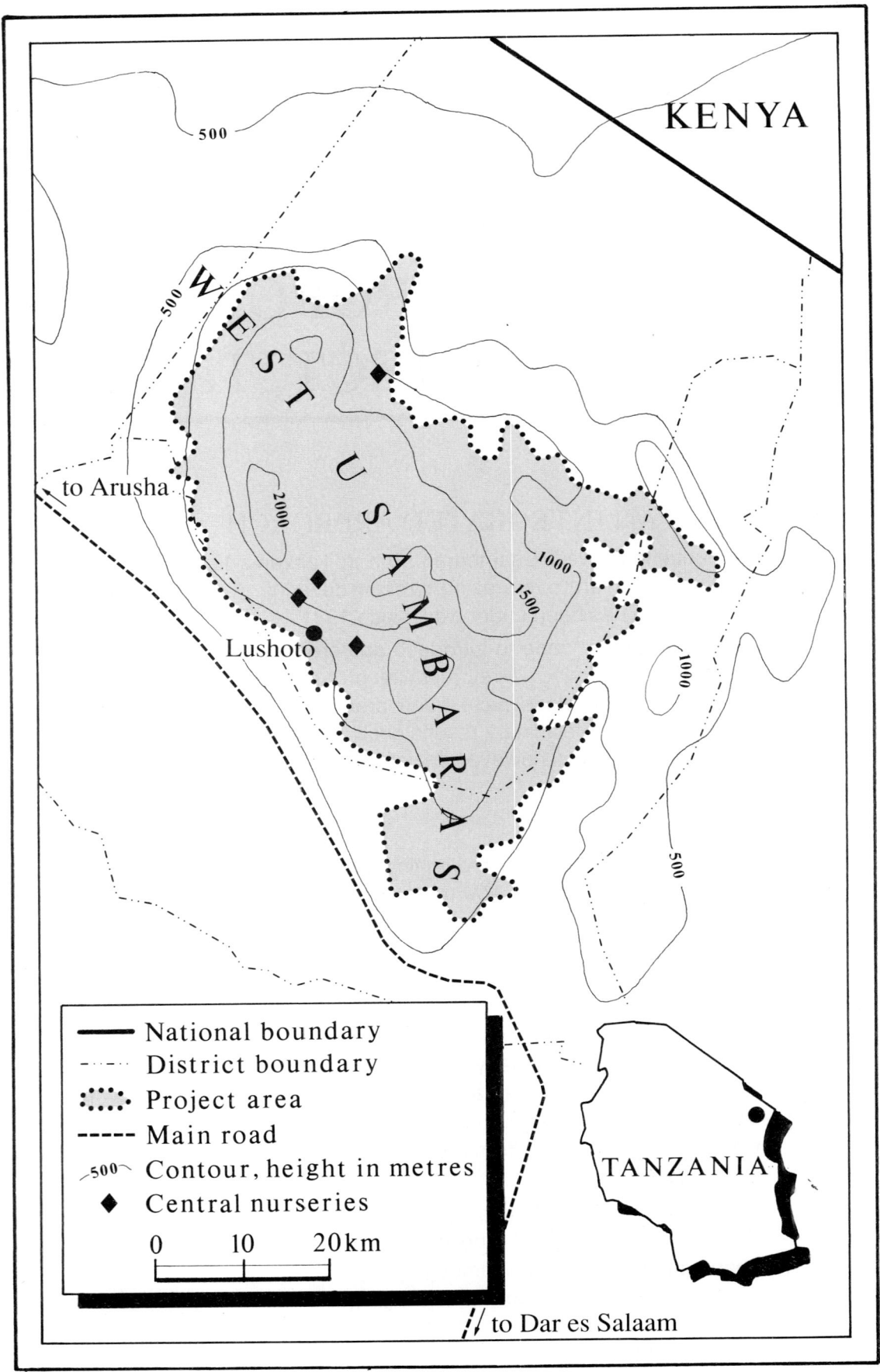

KENYA

to Arusha

W E S T U S A M B A R A S

Lushoto

TANZANIA

to Dar es Salaam

———	National boundary
—·—·—	District boundary
Project area	
– – – –	Main road
⌒500⌒	Contour, height in metres
◆	Central nurseries

0   10   20km

Name of project:	Soil Erosion Control and Agroforestry Project (SECAP)
Address:	P.O.Box 72, Lushoto, Tanzania. Tel: 159.
Project area:	West Usambaras, north-east Tanzania
Average Rainfall:	1,100mm per year in Lushoto
Implementation:	Tanzania Government ministries, with expatriate advisors
Funding:	GTZ US$400,000/year, excluding expatriate costs
	Tanzania Government $30,000/year, excluding staff costs
Exchange rate:	120TSh = 1 US$ (late 1988)

Agricultural activities are divided between men and women, though there is also a considerable overlap in functions. Men tend to produce cash crops in the fertile fields on the valley floors; vegetable production for the Arusha, Tanga and Dar es Salaam markets is particularly important. The upland fields where subsistence crops are grown are more commonly the responsibility of women.

Lars Johansson/Panos Pictures

The cultivation of steep slopes has led to severe erosion in places.

Because of population growth, there has been increasing pressure on land resources over the past few decades. Few, if any, soil protection or conservation measures are used in cultivating the steep uplands. The grazing areas, most of which are on the lower slopes, also suffer from severe over-grazing and soil erosion is rampant.

The project

In 1981, a dairy project was established under the Tanga Integrated Rural Development Programme, with funding from GTZ. In 1984, the project scope was widened to include soil erosion control and agroforestry components and it became known as the Soil Erosion and Agroforestry Project (SECAP). Phase III began in 1988 and has the general goal of the "stabilisation of the ecology in the West Usambaras".

SECAP has an annual budget of about US$400,000 excluding expatriate costs. GTZ is the funding agency and the Government of Tanzania contribution is about TSh4

Contour strips of Guatemala grass are designed to provide fodder and reduce erosion; **grevillea** *trees and some* **leuceana** *have also been planted.*

million (US$30,000) excluding personnel costs.

The dairy farming component is promoting stall-feeding of cattle and the use of cattle manure to increase soil fertility; growing fodder grasses along contour lines was added to support stall feeding. The agroforestry component encourages inter-cropping of trees with food and cash crops and planting trees along contour lines. There is also a forestry component which concentrates on reforesting the hilltops and eroded mountain slopes.

The dairy farming and agroforestry components

The dairy farming and agroforestry components of the project are based upon principles of ecological sustainability and eco-development. All the measures involved are intended to interact and reinforce each other. Most of these eco-development ideas are derived from the Nyabisindu project in Rwanda which is also supported by GTZ.

One of the key ideas is the creation of what are called macro-contour lines. These consist of a strip of grass grown along a contour with a row of trees planted below it and fodder and nitrogen fixing shrubs above. The spacing of the macro-contour lines depends on the slope and varies between 5 and 25 metres.

Among the tree species planted are *Leucaena leucocephela* and *Calliandra calothyrsus*. The shrubs include the perennial creepers *Stylocenthus* and *Desmodium*, both of which the farmers have found difficult to establish. Guatemala grass and elephant grass are among the species used for the grass strips.

The grass strips are intended to provide cattle fodder as well as catching the soil which is presently carried away by rainwater runoff; the hope is that this will eventually lead to the build-up of stable natural terraces. The purpose in growing nitrogen-fixing trees is not just to improve the fertility of the soil; the trees are also intended to cut down erosion and improve the microclimate by reducing temperature extremes. In addition, they provide fuelwood and fodder.

The traditional practice of leaving livestock to graze freely has been made illegal by the local authorities. The intention is that farmers will move to stall-feeding their animals using fodder grown on the contour lines and in special plots. This will increase milk yields as well as enabling the dung to be collected more easily and used as manure. Farmers are also being provided with access to breeding bulls to improve the quality of the stock.

Most of these recommendations have, however, not yet been widely adopted. Stall construction and crossbred cattle are still beyond the reach of the great majority of farmers. Many have simply shifted their cattle down to the lowlands. Others have tied their animals close to their houses where they are fed crop residues, collected fodder and whatever grass the farmers have managed to grow.

One of the more successful village nurseries, managed by local women; about 30,000 seedlings are produced each season.

One obvious reason for the unpopularity of stall feeding is that it requires a lot more labour yet brings a comparatively small improvement in milk yields. A financial analysis carried out by the project shows that, at present prices, the system cannot cover its costs.

There are few data on the numbers of trees or the species planted. One survey shows that farmers are mainly interested in *Grevillea robusta* and other fast-growing species such as eucalyptus. Only 18% of those surveyed grew trees to increase soil fertility and just 2% grew them for cattle fodder. Most of the trees are planted along field boundaries rather than contour lines. This makes considerable sense from the farmer's point of view since it causes less competition with crops.

Experience has also shown that *leucaena* grows very badly. This may be because the climate is too cold, or because the soil is too acidic or too poor. At times, it has been found to grow well when manured, but even then there are some doubts about whether it is fixing nitrogen effectively. Farmers are, however, aware of the indigenous *Albizia schimperiana* which has a solid local reputation for soil improvement and which they will plant if they can obtain seedlings.

Seedling production

Over the first six years of the programme, the total number of seedlings produced amounted to about 2 million. Since the beginning of Phase III, a much greater emphasis has been placed on tree growing and the present production of seedlings is around 1.6 million per year.

The majority of these seedlings are raised in the project's five central nurseries. They are mostly *Grevillea robusta*, *Acrocarpus fraxinifolius*, and *Croton* species and are meant for hilltop reforestation, though some are distributed free of charge to villages if there is a surplus. One of the nurseries is for fruit tree seedlings only.

The project is trying to shift the emphasis away from centralised seedling production. Nurseries have been established in about 80 of the 126 villages in the area. In some cases, the nursery is the direct responsibility of the village, in others it is run by the local school. SECAP provides seeds, polythene bags and technical advice.

In the case of village nurseries, the idea is that the village employs an attendant who looks after the nursery on a daily basis while most of the work is shared among the community. The nurseries each have an estimated average production of 4,000 seedlings per season; this adds up to a total annual production of about 600,000 seedlings.

Despite these impressive-looking production figures, this component is considered to be problematic by the project staff. In many places, people have little trust in the village leadership and are therefore difficult to mobilise for communal work. Moreover, village funds are usually scarce and nursery workers may give up their jobs because they are not paid. Then, when the seedlings in the nursery die, people who participated in the communal work get no return and feel discouraged and disillusioned.

Some project staff now believe that the nurseries need to be further decentralised. They suggest that small groups of farmers, or even individuals, should be encouraged to raise seedlings. In fact, a number of small commercial private nurseries were started in 1988-89 independently of the project. This may well be because, as a result of the project efforts, tree planting is gradually becoming a normal part of local farming life.

The Catchment Forestry Programme

A considerable proportion of the upper mountain slopes in the area have been stripped of their natural forests and used for crop cultivation and grazing. The soils are generally thin and in most cases have deteriorated badly. Soil erosion is widespread and there are fears that the flow of water in streams and wells may be badly affected. In late 1987, an agreement was reached between SECAP and the Ministry of Lands, Natural Resources and Tourism to set up a Catchment Forestry Programme (CFP).

The aim of the CFP is to reafforest the area in order to control erosion and improve water regulation. A high level of village control and management is planned. The intention is that the villagers should be able to obtain timber, poles, fuelwood and fodder up to a limit consistent with the protective function of the forest. In formulating the forest management plan, the aim was to ensure that the village would always obtain a greater return from following the plan than ignoring it.

The management plan is quite different from that normally drawn up in conventional forestry programmes. The villagers are not professional foresters; they are farmers with an interest in maintaining the forest cover to protect their farming environment. A standard plan would be inappropriate since it would leave the villagers dependent on professional foresters and outside funding.

A three-level approach is used. A general management plan concentrates on overall directives and guidelines. Detailed plans for the treatment of each particular area of forest are drawn up by SECAP and village representatives. A village work plan specifies the tasks of the villagers.

The project insists that all the villagers in an area are given a say in drawing up the plans as these will be impossible to implement unless a majority of the local people feel motivated and involved. At the implementation stage, SECAP provides the seedlings, transports them to the village, trains selected villagers in the necessary forestry techniques and normally pays for most of the labour required for planting. Planting the boundary line, of *Eucalyptus saligna*, is a task for the whole village so that everyone knows its exact location.

Once the planting has been done and the boundary established, the area is handed over to the village for protection and management. All the revenues from the sale of produce go to the village. When trees are harvested, in accordance with the plan, the village is responsible for the replanting. Any village which deviates from the plan loses its rights to the area.

The whole process of obtaining village consent to the work and agreeing a management plan is extremely slow. In some cases, it has taken two years of discussion and meetings. It is still too early to make any judgment on the achievements of this component. It nevertheless represents a major effort to systematise village involvement in the management of a large reforestation and forest management programme. The results will provide important lessons for the future, not just in this area but also elsewhere.

Training and extension

Initially, the intention was to use model farms as one of the principal means of extension. Selected farmers were provided with help and inputs so that they could implement the various techniques proposed by SECAP and act as demonstration centres from which other farmers could learn. But the model farms had little spin-off effect and have now been discontinued.

SECAP does not run its own extension service but relies on government extension workers. The Ministry of Agriculture has the largest extension service and is involved in some of the forestry extension such as intercropping of trees in cropland. Much of the forestry extension is done by the foresters.

Extension workers assist in organising village meetings which are normally addressed by divisional or district staff from the Ministry. The meetings are used to bring the attention of farmers to SECAP ideas, such as planting Guatemala grass strips and the use of leguminous shrubs.

The Ministry of Agriculture extension workers have very low salaries, which causes problems with motivation. Many also have little minimal formal education and the senior staff feel that they have very limited professional capabilities. They are not, for instance, allowed to address major village meetings.

School nurseries have been established in an attempt to decentralise seedling production and involve local people.

The project has published a 276-page manual for extension workers in which all major crops and their cultivation techniques are described in detail (Scheinmann and Mchome, 1986). This is available in both English and KiSwahili but is not used much, since the technical terms used are too difficult and it is not always practical; in fact, many of the practices it describes are no longer recommended. It was not tested on extension workers before it was published. Some staff have remarked that it is appreciated by visitors from abroad more than by project staff themselves.

Because of the limited take-up of a number of the project ideas, some of the SECAP senior staff feel that the extension workers need considerably more training in communication skills. A new position has therefore been created in the project for a training and communication officer. Extension aids are also considered necessary but have not been available to date. When they are produced, they may take the form of the coloured pictures on flannelboards used in the Nyabisindu project in Rwanda.

There is, however, a difference of opinion about the need for such additional extension aids. There is a view among some of the project staff that the flaw is not in the extension methods but rather in the package being promoted.

Another matter of discussion is whether the project is damaging to women's interests. The grass strips grown on the slopes, for example, reduce the area of farmland which women cultivate. But the manure obtained from stall-feeding the cattle with the grass grown on these strips goes to the land cultivated by the men on the valley floors rather than back to the women's land.

These are not simple issues to resolve. Soil erosion is no more in the interest of women than it is of men. At the same time, it is important to ensure that the short term and long term social impacts of the erosion control measures are taken into account. Such questions could perhaps be more effectively

Lars Johansson/Panos Pictures

addressed if there were a greater involvement of women in the project.

At present, out of 26 extension workers employed, only four are women; and only two of the 20 subject matter specialists are women. The same imbalance is reflected in the attendance at the village meetings which are the main extension medium used so far: only 25% of those attending are women and the vast majority hardly participate in an active sense. It is, however, difficult for the project to influence the policies of the Tanzanian ministries responsible for the recruitment of their extension staff.

Need to integrate research with project activities

Research into the various components of the project has been carried out at the Forestry Research Station at Lushoto and at several research plots elsewhere in the West Usambaras. Major efforts have been made to assess the effect of macro-contour lines. Species trials have also been carried out to establish the levels of nitrogen fixation and the yields of fuelwood and fodder.

The trials show that the macro-contour lines have a mixed effect on farm production. They occupy 10% of the land and, in addition, reduce the yield of the immediately adjacent crops. In principle, farmers could obtain a compensating increase in yields by applying the manure produced by the stall-fed cattle; but carrying manure to fields on steep slopes is difficult and is rarely done. To set against any such losses, however, there is the contribution of tree products and fodder grass from the macro-contours. They also have a proven capacity to bring about a considerable reduction in soil erosion.

Measurements of the labour involved in creating the macro-contour lines were not carried out during the trials. It is therefore difficult to make a realistic assessment of how much additional work they impose on the women farming the hillsides. In any case, the conditions in a research station are so different from those on a farm that the results of such trials would have been of doubtful relevance.

In the initial species trials, the major focus was on exotics and species such as *Grevillea robusta*, *Acrocarpus fraxinifolius*, *Casuarina cunninghamiana*, *Sesbania sesban*, *Leucaena* species and *Calliandra calothyrsus*. Only one indigenous species, *Albizzia schimperiana*, was tried. In later trials, further indigenous species were included.

The results with indigenous species to date have been discouraging: almost all were found to be very slow growing. *Maesopsis eminii* is an exception though the species is not indigenous to the Usambaras. A continued reliance on exotics therefore seems necessary.

Much of the research has been carried out by PhD students from abroad and has been financed more or less separately from SECAP. This has led to the isolation of the research from the extension work and considerably diminished its practical relevance. It has been noted that the local farmers are sometimes well ahead of researchers yet their knowledge is not being adequately tapped.

There are also feelings among the project staff that the PhD studies are oriented more towards data collection for academic purposes, rather than

31

solving the practical problems in SECAP. Some feel that the research component has had virtually no impact on the programme. This may also be indicative of a lack of communication between researchers and extension staff.

SECAP is fully aware of these problems and intends to strengthen research in several ways. A Tanzanian researcher will be added to the team, and the research component will be institutionally incorporated in the project. It is also agreed that the researchers will have to move from the sanctuary of their research station to the farmer — but nobody yet knows quite how this will be done.

Monitoring and evaluation of project impact

There is a small, separate monitoring unit in SECAP, but project management consider that its staff need considerably more training if they are to become truly effective. At present, some of the more obvious project results such as the number of zero grazing cattle stalls constructed and the quantities of trees leaving the nurseries are measured (see Table 2.1). Monitoring and evaluation of the work carried out by the community nurseries tends, however, to be insufficient.

Nor has much been done to assess the effects of the programme in the field by, for example, trying to establish the numbers of trees planted and their survival rates. All that is known about the macro-contour strips is the quantity of Guatemala grass leaving the nurseries. There is a suspicion among project staff that farmers often leave gaps in the strips. This is counter-productive for soil conservation, but no investigations have yet been carried out. Virtually nothing is known about the impact of the project on farmers' attitudes and agricultural practices.

TABLE 2.1. Monitoring and Evaluation Results, Phase I & II.

OUTPUT	Phase I	Phase II
1. Number of farmers using macro-contour lines	157	n.a.
2. Number of stalls constructed	44	125
3. Number of ox carts sold	n.a.	31
4. Number of fruit trees raised & distributed	n.a.	50,000
5. Area of eroded land rehabilitated (ha)	n.a.	700
6. Amount of Guatemala grass distributed (cubic metres)	n.a.	5,400

References:

JOHANSSON, L. (1988). "Forestry Strategy: SECAP Phase III (1988-1992)". TIRDEP-SECAP, Lushoto.

SCHEINMANN, D. and C.MCHOME (1986). "Caring for the land of the Usambaras. A guide to preserving the environment through agriculture, agroforestry and zero grazing". SECAP.

TAUBE, G. (1988). "An economic analysis of SECAP's basic recommendations for soil erosion control". SECAP.

PROMOTION OF ADAPTED FARMING SYSTEM BASED ON ANIMAL TRACTION, Cameroon

INNOVATION STOPS SHORT OF INTERCROPPING

Traditional farmers in an area of slash and burn agriculture have readily accepted the change to ox ploughs and contour farming. But there has been a much greater reluctance to intercrop with leguminous species. The project shows that farmers are quite prepared to adopt innovations if it can be demonstrated convincingly that they will benefit from them.

Collapse in traditional agriculture

The North-West Province of Cameroon is a mountainous area of about 18,000 square kilometres in which some of the peaks are up to 3,000 metres high. The average rainfall is 2,300mm per year. The natural vegetation is tropical montane rainforest but most of this has disappeared and been replaced by grass and bush.

The estimated average population density in 1988 was 73 persons per square kilometre but in places it reached 150 persons; the population growth rate is in the range 2.4-3.0% per annum. There is a large amount of ethnic diversity with languages changing over intervals as short as 20 kilometres. Pidgin English was introduced at the beginning of the century as the lingua franca, and now serves as a generally understood language among the many ethnic groups.

Farmers have traditionally relied on slash and burn agriculture. The vegetation in an area is cut, covered with earth and set on fire; it is roughly the same process as used in charcoal making. It is known locally as "bury and burn" and results in a good yield for the first three years of cultivation.

With the increase in population density, fallow periods are becoming so short that the bury and burn method has become impossible to use in many areas. Instead, farmers are turning to permanent agriculture. But either through ignorance of the appropriate techniques or lack of resources, they are generally unable to maintain the fertility of the soil. The result is widespread land degradation and falling agricultural yields.

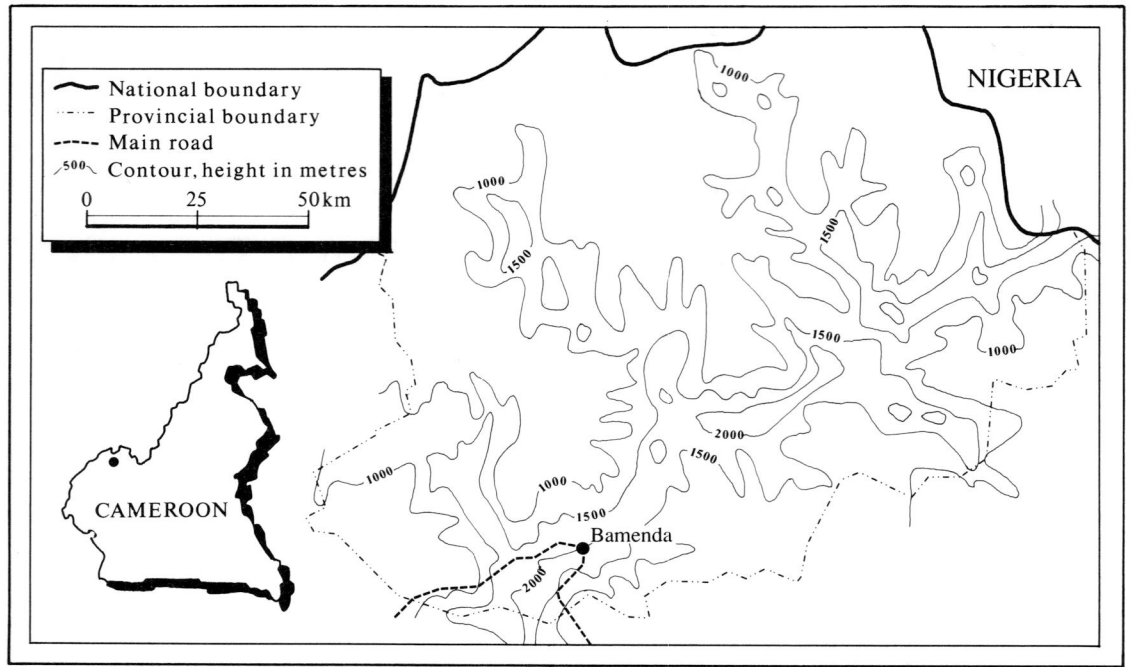

Name of project:	Promotion of adapted farming systems based on animal traction (PAFSAT)
Address:	BP 558, Bamenda, Cameroon. Tel: 362510, Tlx: 5087KN
Project area:	North West Province
Average rainfall:	2,300mm per year
Implementation:	GTZ and Ministry of Agriculture
Funding:1984-present:	US$ 270-800,000 per year
Exchange rate:	300 FCFA = 1 US$ (early 1989)

The PAFSAT Project

In the late 1960s, GTZ launched an agricultural mechanisation programme. The aim was to reduce the burden of hoe cultivation and allow farmers to cultivate larger areas as a means of countering the fall in crop yields which was even then taking place.

These objectives were to be achieved by introducing tractors for ploughing and other farming tasks. This was not a success and the project changed its approach to the promotion of animal traction and improved agricultural techniques. The animal traction component became an independent project in the North-West Province in 1980.

It began as a pilot project and concentrated on developing animal traction techniques for different farming operations; it also carried out research into appropriate permanent farming methods including erosion control and the maintenance and improvement of soil fertility. In 1984, the PAFSAT project was launched and has been running since with an annual budget which has varied from US$270,000 to US$800,000.

1. Nocturnal stable keeping of oxen to produce manure.
2. Erosion control by contour bunds reinforced with permanent crops.
3. Mixed cropping and crop rotation for soil improvement and pest control.
4. Planting of annual crops on humus rich field ridges.
5. Row intercropping with legumes instead of sole cropping.
6. Preparation of ridges and weeding by oxen with the ridger plough.
7. Integration of formerly separate planted cash and food crops in one field.

Diagram of the PAFSAT farming system.

The aim of the project is to promote permanent farming based on the use of oxen. The target groups are crop farmers and livestock herders. A special emphasis is laid on the participation of women, as members of farming families, as independent farmers, or in women's farming groups.

Under the project, farmers are given a six-week training course in animal traction and permanent farming techniques. They are supplied with animals and equipment, such as ploughs and carts, on five years' credit. The project's extension service provides follow-up advice on farming methods, animal husbandry, and the care and maintenance of equipment.

The emphasis in the farming methods being promoted is upon erosion control and the maintenance of soil fertility. Among the recommended measures are the use of contour bunds reinforced with permanent crops, contour farming of seasonal crops, use of cowdung and fertiliser, manuring with crop residues and mixed cropping.

The use of green manure crops is also promoted. These include *Sesbania sesban*, *Tephrosia vogelii* and *Crotalaria*. It is recommended that they are planted on the contour bunds at the same time as food crops or that they are used as part of a planted fallow system. They are cut after three months to a year, and the leaves and stems are worked into the ground. *Leucaena leucocephala* is not used because it has been found to be too slow growing.

The project, after an initial period in which it concentrated on male-headed families, has now integrated women into its training schemes. Considerably more women than men are now participating, and unlike many other parts of Africa, they are often found behind the plough (Linz and Zalang, 1988).

A major on-farm research programme

The project runs a major on-farm research programme to support and develop its recommended innovations. From 1985 onwards, a variety of trials have been carried out on about 80 farms selected to provide a representative cross-section

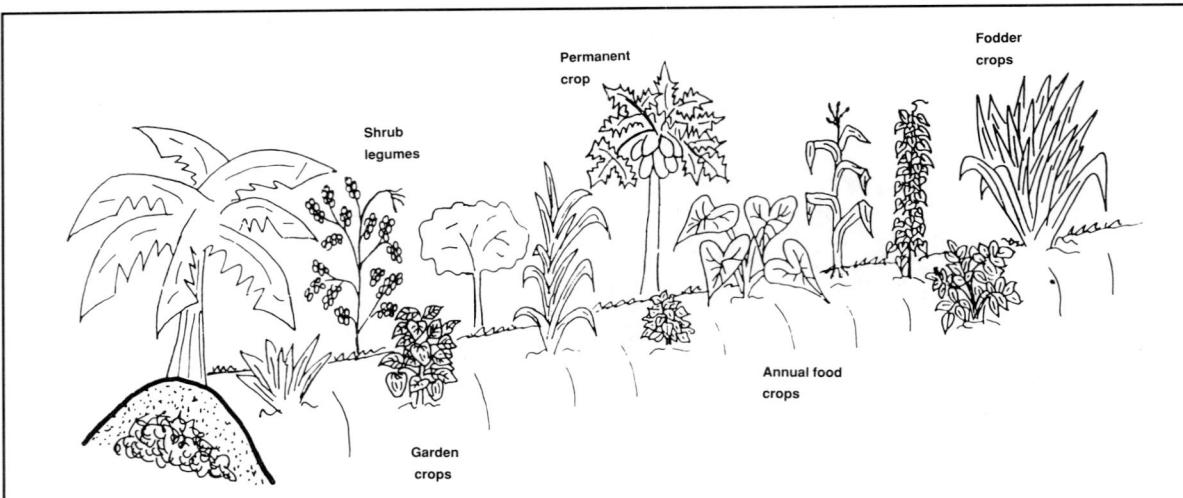

Shrub legumes

Permanent crop

Fodder crops

Annual food crops

Garden crops

The contour bund protects your soil against erosion and contributes to your farm income.

Apart from permanent crops which guarantee a long-term income, you can plant seasonal crops like: beans, soya-beans, groundnuts, corn, bitter leaves, pepper, okra, etc., for fast income.

Diagram from the PAFSAT farmers' handbook describing benefits of contour bunds.

of soils, altitudes and micro-climates.

The trial plots include the basic elements of the recommended farming system. The layout consists of five contour bunds planted with perennial crops such as plantain, coffee and legumes for reinforcement, intercropped with seasonal crops such as beans, groundnuts and soybeans for immediate production. The strips of field between the bunds are used to demonstrate different crops and crop rotations. The objective is to provide farmers with practical examples which they can either copy or adapt.

The effects of different treatments on crop yields and soil properties are also being examined in the trials. Among the treatments are the use of cowdung and fertiliser and intercropping maize with leguminous species. The intention is that soil analyses will be carried out every four years to establish the effects of the different treatments.

In the initial stage of the trial programme, the extension staff of the project supervised and directed the trials closely. The farmer, without any real appreciation of what was happening, was reduced more or less to a bystander, simply carrying out the tasks allocated by the project staff. The plot was the "PAFSAT plot" and the trial was "the responsibility of PAFSAT".

It became clear that this was totally unsatisfactory. The trial plots require various treatments such as manuring and weeding; they are also subjected to unexpected disturbances like hail and intruding livestock. Only the farmer, working on his own crops beside the trial plot, is able to carry out the necessary treatments, make the required observations and note the disturbances and distorting factors. Project staff cannot provide this degree of continuous supervision; moreover, they have a variety of other tasks which they must perform.

The participation of farmers in their own trials has therefore been increased.

During a two-day seminar with trial farmers and staff, the bottlenecks in the process were discussed and the various responsibilities were newly defined. The farmers concerned have been provided with information on the trial objectives and received training on the establishment and maintenance of the plots. They are now responsible for the running of the trials including making the relevant measurements at harvest and other times.

Supervision, guidance and assistance are provided by the project using one senior full-time staff member for planning, co-ordination and evaluation, and extension workers for the field follow-up. Each extension worker is responsible for overseeing 3-4 trial farms.

Trial results to date

So far, only about 50% of the trials have produced usable data; the rest have been disturbed or distorted in a variety of ways. The fact that there is still a considerable number of reliable results means, however, that a reasonable degree of confidence can be placed on the findings.

One clear conclusion is that major benefits are obtained from the use of cowdung or fertiliser. The trials show that yields are increased by 50% with 10 tonnes of cowdung or 250 kilogram of NPK fertiliser (20:10:10) per hectare and by 100% if both are used.

Preliminary results show that the traditional "bury and burn" system has a higher productivity in the first year than the method recommended by PAFSAT. The results from the four-year trials are still to come but are expected to show a decline in the traditional system compared with more or less constant yields from the PAFSAT method.

Widespread adoption of new techniques

The farmers have readily adopted new techniques such as ox-ploughing and the use of contour ridges. There has also been a widespread acceptance of

the use of cowdung, and the recycling of crop residues. This appears to be a direct result of their demonstration in the on-farm trials. The use of comparative treatments, such as the application of dung and fertiliser, which allow farmers to draw their own conclusions from observation of the results, have been particularly effective.

Intercropping food crops with legumes has, however, been poorly received by the farmers. They regard the use of the legumes as a loss of productive land and are not convinced of the longer term benefits. It is also clear from the trials that more information is required on the optimum planting times and intercropping arrangements to avoid reductions in the food yields. These questions are being investigated in a series of on-station trials which were launched in 1989.

The project has obviously succeeded in a number of its basic objectives. It clearly shows that farmers are quite prepared to adopt innovations which have demonstrated their effectiveness and relevance. But, as has been found in a host of other projects in Africa, they will not adopt new measures if they are not convinced they will benefit from them, no matter how strongly these are promoted by project staff.

References:

LINZ, S. and F.MPOL ZALANG, (1988). "Impact of the PAFSAT project on the situation of the women". PAFSAT/GTZ.

PAFSAT (1988). "On-farm observation and demonstration trials". Technical handout.

ZWEIER, K. (1988). "PAFSAT Farmers Handbook". PAFSAT/GTZ.

THE GITUZA FORESTRY PROJECT, Rwanda

FUEL FOR REFUGEES

Substantial progress has been made in planting trees for fuelwood and promoting agroforestry in Gituza Commune in northern Rwanda. Large parts of the area are now covered in plantations and a variety of new exotic tree species have been introduced on most farms.

The programme has shown what can be accomplished given an effective implementation organisation and a high level of funding concentrated in a small area. The question is whether the achievements are sustainable in the long term.

The project

The project area is Gituza Commune, an administrative unit of 360 square kilometres, situated close to the border with Uganda. It is a region of rolling hills and valleys with altitudes ranging from 1,400 up to 1,900 metres. The climate is relatively dry for Rwanda with an annual rainfall of about 1,000mm. The colonisation of this part of Rwanda by farmers is quite recent.

Some 210 square kilometres of the area are within the Akagera National Park. The remainder consists of densely populated agricultural land with a total population of 33,000 people. Subsistence crops are cultivated in the valleys, and the hilltops are used for communal grazing.

During the civil war in Uganda, some 23,000 refugees arrived and were settled in a camp near the national park. They relied on the neighbouring woodlands and wood bought from local Rwandan villagers by the UN High Commission for Refugees (UNHCR) for their fuel supplies.

The project was designed in 1984 and was primarily oriented towards the fuelwood problems resulting from the influx of refugees. It included forestry plantations for fuelwood production; the promotion of agroforestry to increase fuelwood supplies and relieve pressure on existing resources; and the

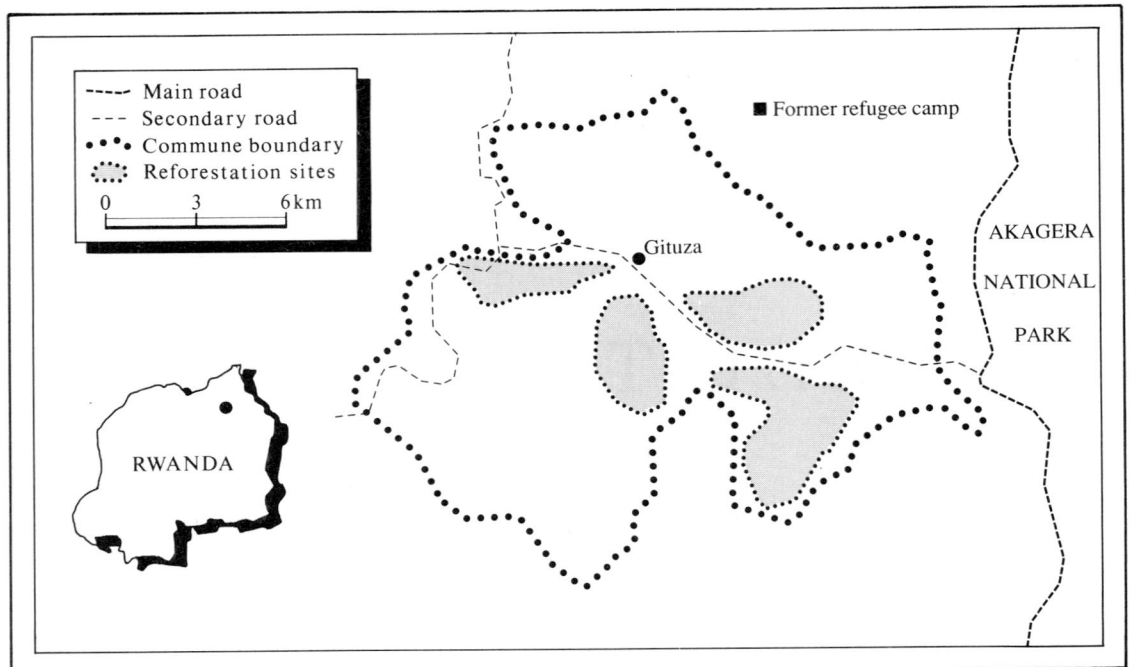

Name of project:	**Gituza Forestry Project**
Address:	**c/o CARE (Rwanda), BP 550, Kigali, Rwanda**
Project area:	**Gituza commune**
Average rainfall:	**1,000 mm per year**
Implementation:	**CARE (Rwanda)**
Funding:	**1985-89: US$1 million/year (USAID)**
	1989-93: (projected) US$1.1 million/year (DGIS)
Exchange rate:	**80 Rwanda Francs = 1 US$ (October 1989)**

introduction of improved wood and charcoal stoves.

The expatriate project manager and the Rwandan assistant manager are employed by CARE, a US-based NGO. Others are employed by Minagri, the ministry co-ordinating the Departments of Agriculture, Forestry, and Livestock. Extension workers are initially employed by CARE but then move to Minagri. Nursery staff are normally paid by CARE.

The fuelwood component

The initial project document estimated that 3,600 hectares of plantation would be required to meet the fuelwood needs of the refugees. This was based on an assumed consumption of 2 cubic metres per person per year.

The land designated for the plantation consisted of hilltop communal grazing areas, some of which had been damaged by fires and showed signs of erosion. Village meetings were called to explain the project and, thereafter, no grazing or other activities were allowed in the planting areas. Some of the proposed

Rudolph von Bernuth/CARE

Pine plantation on one of the hills surrounding Gituza; the trees are growing well but the future of the plantations is uncertain.

boundaries were, however, altered after consultation with local people.

All the labour required for the planting was provided by the project, largely through a food-for-work programme funded by UNHCR. Micro-terraces were built on the planting sites, chalk was added to acid soils and pesticides were used for protection against termites. The main tree species were *Pinus, Callitris* and *Acacia mearnsii.*

Watchmen were provided at the rate of one per 50 hectares and increased to two per 50 hectares in the dry season. The establishment of the plantations was generally very successful. The older ones already give the visual impression of a closed forest and seem to be doing better than those of a nearby project funded by the World Bank. The total area planted under the project in Gituza Commune is 1,850 hectares and a further 650 hectares have been established in a neighbouring commune. About 80% are classical conifer plantations and the rest are of broad-leaved species. After 1987, no more plantations were established.

In 1986, a year after the project began, the refugees returned to Uganda. The major funding agency, USAID, decided it would continue funding the project, but only for the first 3-year phase which ended in 1988. The stated objectives of the programme were changed from "fuelwood production" to "soil conservation" but the strategy remained essentially unaltered.

Questions over the future of the fuelwood plantations

The plantations are presently managed by the project staff in collaboration with the Forest Department. The major costs are borne by the project.

The 1984 project document mentions that "10% of the proceeds of the plantation go to the commune", but no detailed plans for the distribution of benefits have yet been drawn up. The local people have, in fact, had little involvement with the project except as paid labour.

The costs of managing and protecting 2,500 hectares of newly established forest plantations are high. They include thinning, pruning, and the provision of watchmen, and will have to be borne by the Forest Department if the project funding ends. It is unlikely, given the low incomes in the area, that nearby markets will be found to recover any significant proportion of the costs by means of fuelwood or timber sales.

The departure of the refugees has, in fact, left the forestry component of the project in something of a vacuum. While they were there, it could be justified on the grounds that it was supplying fuelwood needs. The purchase of wood for this purpose by the UNHCR also provided a cash return which could cover at least some of the running costs of the plantations.

Simply changing the nominal objectives of the programme from fuelwood production to soil conservation does not resolve the problem. Soil conservation strategies are normally different from those for fuelwood production. The *Pinus* species, for example, do not have a particularly good conservation reputation. In general, plantation forestry is an expensive and not necessarily effective method of soil conservation.

It is true that the construction of micro-terraces is a proven remedial measure on badly eroded slopes but this is not specifically related to the planting of fuelwood or indeed any trees. The assistant manager of the project explained that, as a result of the reforestation, the area is now better respected by the local people and they are less likely to burn it.

A recent evaluation pointed out the high costs of plantation management. It suggested reducing these by using local "voluntary labour" and selling the wood products in the capital, Kigali. This, however, appears questionable, as it could well result in exploiting Gituza inhabitants to subsidise the city. Such discussions about the plantations highlight their precarious future (Speich et al, 1988).

The agroforestry component

A diagnostic survey of the area was carried out in 1985, soon after the beginning of the project. This also provided baseline data against which subsequent achievements could be measured.

The survey used a questionnaire and covered a variety of topics such as which crops were cultivated, the numbers of livestock kept, attitudes to tree planting, and the problems experienced by farmers. It was co-ordinated by senior CARE personnel but was carried out by locally recruited project staff. These subsequently became extension workers when the project was under way. The survey helped increase the local knowledge and understanding of the project staff at all levels and hence had a training as well as a planning function.

The survey found that poor soil fertility was the major constraint on food production in the area. It also revealed that tree growing was common among farmers, with eucalyptus species for timber and *Euphorbia tirucalli* for hedges being the most commonly planted. The trees are grown in a variety of configurations, with about 50% planted in compounds and the remainder divided between croplands and hedges. The survey results also showed that fuelwood shortages were not seen as a major problem by the majority of farmers.

As a result of these findings, it was decided that high yielding nitrogen fixing trees provided the best response to the soil fertility constraints which weighed heavily on farmers. As there was little experience with such trees in the area, some research was considered necessary. This took the form of a trial plot in each of the two ecological zones of Gituza.

The research was carried out by CARE staff and mainly concentrated on trials of exotic species such as *Leucaena leucocephala*, *Calliandra calothyrsus* and *Gliricidia sepium*, and the indigenous *Sesbania sesban*. These were planted in

Rudolph von Bernuth/CARE

The central nursery near Gituza village.

hedges in conjunction with food crops, and were managed with alley-cropping techniques. Some fruit trees were also included in the trials.

The research results to date have been processed by computer by CARE staff in Kigali. They include data on the wood and foliage yields of trees but no firm data on the interaction between trees and crops have yet been obtained.

Pending the result of the trials, a package of measures was adopted for promotion. This consisted mainly of intercropping leguminous shrubs in foodcrops; promoting tree species such as *Markhamia platycalyx* and *Cedrela fissilis*; planting the same species in association with grass strips along contour lines; and planting fruit trees near homesteads.

Seedling production and distribution

The Government of Rwanda has enforced a policy whereby all able-bodied people are required to spend one day per week working for community projects, the so-called "Umuganda". Some of the Umuganda is to be used for a tree nursery in the commune. In theory, this provides an extensive seedling production system throughout the country, with free seedlings available to all Rwandan farmers.

Project staff were, however, aware that the Umuganda nurseries did not function particularly well. They therefore proposed that the project nurseries should be run with paid labour and that the seedlings should be sold at a subsidised price rather than given away free. Minagri, which has responsibility for the Umuganda nurseries, accepted the proposal as an experiment.

The project has one large central nursery which serves both the forestry and agroforestry components. It formerly had a capacity of about 1 million seedlings per year, but this has been reduced since the establishment of new fuelwood plantations stopped. A series of smaller nurseries have also been established in the subdivisions of the commune. These are run by the local extension worker with the aid of some casual labour and have a capacity of about 10,000 seedlings each per year. The extension workers are expected to find out from the farmers the number of seedlings required in their subdivision; the farmers collect the seedlings themselves.

Staff training and extension

When the country gained its independence in 1962, there was a major shortage of trained personnel. Lack of skilled local staff still remains a major problem and the eight extension workers recruited by the project required a considerable amount of training.

The work on the diagnostic survey provided a preliminary grounding in that it involved them in discussions on agroforestry with a considerable number of farmers. They were then given several months' training in communication skills and subjects such as the importance of trees in the environment and the selection of appropriate species and tree planting sites.

Even then, it was found that they remained far from effective because the farmers expected a much wider range of skills from them. If, for example, the

44

farmer being visited had a diseased tomato crop, he expected help with that before discussing fuelwood or agroforestry. The reason for this is that the extension service responsible for such crops is severely understaffed and has few resources. There are, in fact, only 10 agricultural extension workers in the whole of Gituza Commune.

The project therefore decided that its extension workers needed a training in general agricultural subjects. This is provided in the form of part-time courses by INADES, a Francophone organisation specialised in rural development training.

The project area is divided between the eight extension workers. They are also responsible for the nursery in their area. In promoting agroforestry, they use two main techniques which the project refers to as "intensive extension" and "extensive extension". In addition, there is a demonstration plot at each local nursery.

The intensive method relies on visiting a small number of farmers regularly throughout a season. During the following year a different group is visited.

The extensive method uses village meetings. In these, an extension worker makes a presentation on a particular theme such as erosion control or fuelwood. Flannel boards with drawings are used to increase interest and help provoke discussion, using the extension model of the PAP Project in Nyabisindu, Rwanda. Farmers who participate in these meetings are given a discount on the price of seedlings.

A recent evaluation concluded, however, that most extension workers are not able to stimulate discussion. It was also found that farmers get bored with the drawings, and it was recommended that another type of audio-visual aid should be developed (Speich et al, 1988).

Table 4.1. Data on impact of the agroforestry component

	Before project	Second season after commencement
Percentage of farmers who have planted trees in cropland	42%	83%
Percentage of farmers who have planted trees on boundaries	61%	82%
Percentage of farmers who have planted:		
Eucalyptus spp:	80%	37%
Leucaena leucocephala:	0%	20%
Cedrela fissilis:	0%	23%

Source: Gibson and Muller (1987)

Monitoring and evaluation

The project has developed a strong monitoring and evaluation system which enables it to assess its effectiveness and change strategies accordingly. It relies on three basic techniques: a baseline survey, seedling distribution lists, and follow-up visits to the farms six months after the seedlings have been delivered. The visits are not just to check the rate of seedling survival; they are also intended to discover the farmer's reaction to new species, find out how the trees are used, and make an estimate of the likely demand for seedlings the following year.

The results to date show that the project has had a considerable impact. The number of trees planted in croplands, on boundaries, and along contour strips has increased markedly. There has also been a switch from eucalyptus to nitrogen fixing species such as *Leucaena* and *Cedrela. fissilis*. There are, however, no data yet to show whether alley cropping *Leucaena* with food crops has produced any significant increase in yields or overall benefits to farmers. Table 4.1 shows some of the results obtained to date.

The main problem with the present monitoring and evaluation system is that it requires large amounts of skilled and costly manpower; it also relies on computer processing in Kigali. The baseline survey, for example, needed six months' external consultancy. Heavy demands are also made on the project staff. The system in its current form may therefore not be replicable where high levels of external funding and expertise are not available.

The stoves component

In the beginning, the project planned to reduce charcoal consumption by introducing improved stoves. This was abandoned when it was found that charcoal was not used in the project area.

A ceramic woodstove called the "Zigamingkwi" (meaning "save a lot") was then developed by the project and is being sold at a subsidised price of about 150 Rwandan Francs (about US$2.00). Sales figures are low. In April and May 1988, for example, a total of only 25 stoves per month were sold.

This element of the project is not having any significant impact and staff regard it mainly as a consciousness-raising exercise. The stove design is not satisfactory and many crack within a short period of use. One survey shows that 50% are abandoned because of breakage or for other reasons.

Sustainability and replicability

The fuelwood component of the project was targeted on the highly mobile and unpredictable refugee population. The refugees left, and as a result, 12% of Gituza Commune is now covered in fuelwood plantations for which no commercial demand can be expected. This demonstrates the danger of using long term interventions, such as reforestation, in response to variable or short term problems such as an influx of refugees. A change in circumstances can easily make the response irrelevant or even turn it into a liability for the future. In such cases, it is far better to devise short term and flexible solutions which can be adjusted as necessary if the situation changes.

It is fortunate that soil conservation of the hills is served by the afforestation which has taken place. On the other hand, this cannot be considered a model for the conservation of the Rwandan hills. The 2,500 hectares of forest which have been established will, almost certainly, need continued external assistance to safeguard the investment, since the Forest Department does not have the necessary resources.

The techniques developed in the agroforestry component are, however, likely to find application elsewhere. The method developed for seedling production and distribution, for example, offers an alternative to the present rather ineffective Umuganda system. The monitoring and evaluation system is a major improvement on that in many other forestry projects where no one really knows what is happening to the seedlings and what impact the extension work is having. The fact that it relies upon data processing by computer will, however, restrict its application to projects which are comparatively well endowed with funds.

A major worry, however, is that the level of support for the agroforestry activities is far greater than can be maintained after the project ends. At present, there is an agroforestry extension worker for every agricultural extension worker. The project staff fear that, when the funds dry up, most of the senior and junior staff working on the project will be transferred elsewhere. In the absence of the CARE staff, that will mean the collapse of the nursery system.

It may, therefore, turn out that, while the programme has undoubtedly had a significant effect on local tree growing patterns, this has been principally a result of its high level of funding. Unless the farmers find they are obtaining readily apparent increases in crop yields and other immediately tangible benefits, they may abandon the new practices once there is no pressure from extension workers to keep them up. This will be a particularly serious danger if the necessary seedlings become difficult to obtain.

References:

GIBSON, D.C. and E.U. MULLER (1987). "Diagnostic surveys and management information systems in agroforestry project implementation - a case study from Rwanda". ICRAF Working Paper No.49.

MULLER, E.U. (1987). "Draft case study of the CARE Gituza Forestry Project in Rwanda". Agroforestry Monitoring and Evaluation Project.

SPEICH, A. et.al. (1988). "Gituza Forestry Project Evaluation".

KENYA WOODFUEL DEVELOPMENT PROGRAMME

CONFRONTING 'THE OTHER ENERGY CRISIS'

A major programme to develop effective methods of promoting tree growing for fuelwood has been under way in the Kenya Highlands for the past five years.

The achievements of the programme are substantial. In particular, it has made major strides in developing rural survey techniques. But questions are being asked about whether it has truly focused on the real tree growing priorities of the rural people involved. There are also doubts whether some of the extension packages it has developed can be adopted by the existing government extension services when project funding ends.

Project origins

During the late 1970s there was a rising tide of concern about "the other energy crisis": the depletion of the fuelwood resources on which the majority of rural people in the Third World depend for their energy needs. In Kenya, a large study carried out for the Ministry of Energy in 1979-82 predicted severe fuelwood shortages, particularly in the densely populated Highlands of the country.

The Kenya Woodfuel Development Programme (KWDP) was one of several woodfuel projects designed to deal with these impending fuelwood problems. It is under the supervision of the Ministry of Energy and began in 1983. Its principal objectives are to develop the tree cultivation techniques and extension methods required to promote increased fuelwood production by farmers in the Highlands; and, in the longer term, to hand over the running of the programme to the various government ministries concerned.

The programme is carried out in two districts, Kakamega and Kisii, both in the Highlands. Most of the area covered is at an altitude of about 1,500 metres and has a rainfall of 1,600-2,000mm per year. It is one of the most fertile parts of the country and is officially classed as having a high agricultural potential.

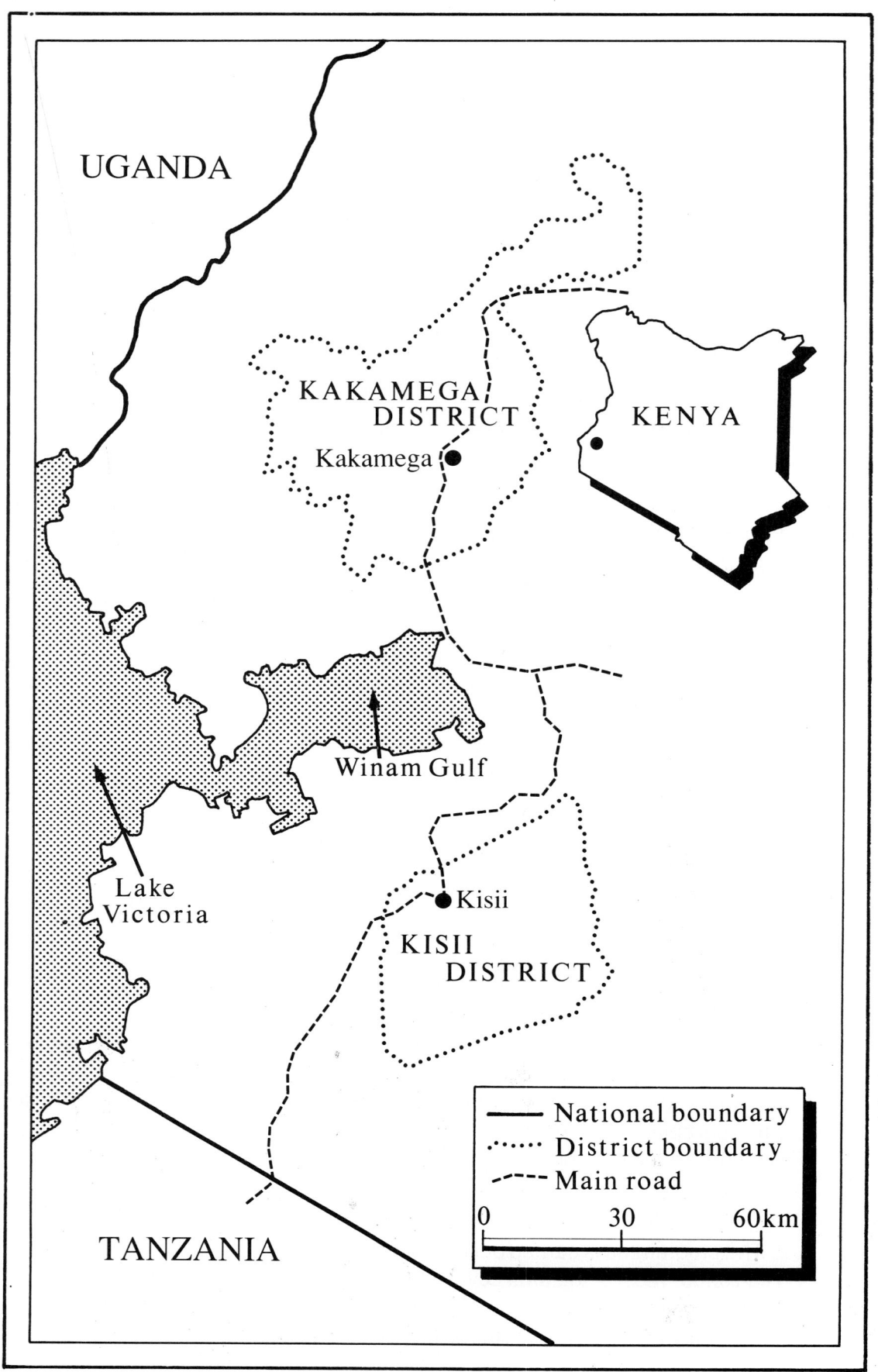

UGANDA

KENYA

KAKAMEGA
DISTRICT

Kakamega ●

Winam Gulf

Lake
Victoria

Kisii ●

KISII
DISTRICT

—— National boundary
······ District boundary
- - - Main road

0   30   60km

TANZANIA

Name of project:	Kenya Woodfuel Development Programme (KWDP) Now called the Kenya Woodfuel and Agroforestry Project (KWAP)
Address:	1. P.O.Box 1080, Kakamega, Kenya. Tel: 0331 20456 2. P.O.Box 2417, Kisii, Kenya. Tel: 0381 20094
Project area:	Kakamega and Kisii Districts
Average rainfall:	1,500-2,000 mm per year
Implementation:	1983-88 Beijer Institute, Stockholm, in conjunction with the ETC Foundation, Netherlands
1989 - present:	ETC Foundation.
Funding:	1983-88 approximately $10 million (DGIS)
Exchange rate:	18 KSh = $1.0 (early 1989)

Each of the two districts has about 1.25 million people. The population density is extremely high and varies from around 200 to as many as 1,000 people per square kilometre in some places. The rate of population growth is 4% per year. Land holdings are generally small, with some families owning only 0.2 hectares. There is little forest or communal land.

Funding for the programme has been provided by the Netherlands Government. Until recently, the KWDP had a project office in Kakamega and Kisii as well as headquarters in Nairobi; but the Nairobi office has now been closed. In 1988 there were doubts about future funding, which hampered operations, but from April 1989 project activities were resumed.

Baseline surveys

After the programme was set up, and before any decision was made on the techniques to be used to promote tree growing, extensive preparatory surveys were carried out in both Kakamega and Kisii. Three different approaches were used.

Chris Pennarts/Panos Pictures

The "District Resource Analysis" employed air and satellite photographs as well as a certain amount of field-work. It developed a land classification system based on 30 characteristics such as farm size, type of crops, extent of hedgerows and woodlots. Detailed maps were then produced showing the land in each district divided into 11 zones in accordance with the

Tree growing is a well established practice in Kisii and Kakamega districts; the more dense the population, the more trees are generally grown.

classification system.

This analysis showed that indigenous trees and shrubs tend to be replaced with exotic species, especially eucalyptus, as the population density grows. It also produced the rather surprising information that the proportion of land with planted trees increases with population density.

Jan Kuyper

In areas with a relatively low population density, say 200 people per square kilometre, the land is largely under cultivation though there is still some bush left, and few trees *Simple 'farm nurseries' were found on many farms, proving that farmers already know how to raise seedlings; this led to a major shift in project thinking.*

have been planted. In areas of high population density, say over 500 people per square kilometre, farmers have planted a large number of trees on their farms, covering up to 30% of their land (Bradley, 1988).

The second set of surveys covered agroforestry. The aim was to establish what techniques farmers employed for tree propagation, and how they managed and used their trees. About 500 farmers in Kakamega and 250 in Kisii were interviewed. They were asked whether they used cuttings, direct seeding, or relied on tree nurseries. They were also asked what use they made of the trees.

These surveys revealed that tree growing is extremely common. Hedges, woodlots and windbreaks are a feature of most farms, and trees are normally also found on farmlands and in the compound. The patterns of tree growing, however, vary considerably between and within the different districts (van Gelder and Kerkhof, 1984; Kuyper and Bradley, 1985).

The surveys also showed that farmers are much more involved in tree propagation than had been expected and are quite capable of producing their own seedlings. It was found that "micro nurseries" exist on about a third of all farms, with over 50,000 in Kakamega District alone. The vast majority of the trees raised in such nurseries are construction wood species like eucalyptus and *Cupressus lusitanica*.

Most farmers establish their nursery at the onset of the rainy season. It is often located in a protected and shaded place like the banana grove just behind the house. The seedlings are raised without using pots and may be planted out during the same rainy season. Survival after transplanting is quite low, but so also is the investment made by the farmer.

The realisation that farmers are able to produce their own seedlings changed the initial project strategy drastically. In the project document it had been assumed that farmers got their seedlings from the Forest Department and that the absence of a nearby nursery would prevent them planting trees. The project had therefore proposed setting up a system of decentralised community nurseries. When the information from the agroforestry survey became available, there was a shift in the project focus to the farm nurseries.

The third set of surveys was concerned with cultural constraints on increased woodfuel production (Chavangi, 1984; Ong'ayo, 1986). These surveys were based on in-depth interviews of women's groups and community groups by senior project staff; individual people were also interviewed. Since women are responsible for fuelwood collection and men own the land, much of the interview focus was on women's access to men's resources, and how conflicts are resolved.

The cultural surveys showed that there were "men's species" like *Cupressus* and eucalyptus which produced construction poles, and "women's species" such as *Sesbania sesban* which are not really suitable for construction and are mainly used for cooking fuel. In Kakamega, for example, women are not allowed to plant eucalyptus, but are free to grow and harvest *Sesbania*.

The cultural surveys revealed that fuelwood is not discussed between men and women and that few, if any, farm resources, other than women's labour, are allocated to its production. At the same time, fuelwood collection involves considerable drudgery and many families are often forced to buy some.

It was also found that there are severe obstacles to tree planting by women. Some are prevented by their husbands; there are also some taboos such as the belief that "if a woman plants a tree, she will become barren." As a result of these surveys, the KWDP concluded that the cultural constraints on growing trees for fuelwood were the main problem it had to face.

Developing a credible technical package

To promote increased tree growing for fuelwood it was necessary to develop a technical package which was credible to local farmers. Species trials were therefore started at the existing forestry research stations in each district; in addition, a number of on-farm trials were begun.

The research station trials covered species selection and such variables as spacing and management. In Kakamega, the trees were planted as woodlots using three different spacings. In Kisii, hedgerows were tested since the surveys showed that farmers often planted trees in hedges. The main species used in the

Chris Pennarts/Panos Pictures

*The farmer has planted **Calliandra calothyrsus** in maize; KWDP researchers left farmers to design their own on-farm trials.*

Table 5.1. Results from fuelwood production trials in Kakamega and Kisii Districts

KAKAMEGA:

Woodlots established in 1985, yields after one year growth (fresh weight in tonnes per hectare).

		spacing	
Species	0.5x0.5m	1.0x1.0m	1.5x1.5m
Calliandra calothyrsus	62	34	35
Sesbania sesban	81	57	46
Mimosa scabrella	38	36	34

KISII:

Hedgerow trials established in 1986, yields in kg/m of hedge one year after establishment (Kilograms per metre of hedge)

Species:	Spacing:	Yield:
Sesbania sesban	0.2x0.1m	17 kg/m
Sesbania bispinosa	0.2x0.1m	19 kg/m
Mimosa scabrella	0.3x0.3m	16 kg/m
Calliandra calothyrsus		5 kg/m
Leucaena leucocephala		3 kg/m

Source: Mbuthia (1987)

---

hedgerow trials were *Leucaena leucocephala*, *Calliandra calothyrsus*, *Gliricidia sepium*, *Mimosa scabrella*, *Sesbania* species, and *Markhamia lutea*.

The trees were cut after one year or more to study the subsequent production from the coppice. The full biomass production, that is wood plus leaves, was measured in each case (see Table 5.1). The major conclusion of the trials was that the fuelwood production capacity of woodlots and hedges is very high in the two districts. The data from Kisii show that a hedge of *Sesbania sesban* around a farm of 1 hectare would produce perhaps half the fuelwood requirements of an average family.

Eucalyptus was not included in the trials although it is a well known tree in both districts and widely appreciated by farmers. This decision was partly because it is not seen as a fuelwood species; it was also because it is well-established in the area and its performance is therefore well known. It would, nevertheless, have been useful to see how it compared with some of the new exotics under trial conditions.

A series of on-farm trials was also carried out. The farmers involved were offered seedlings and seeds of the new exotic species, as well as some indigenous trees, and allowed to select which they preferred. They also planted the trees in whatever way they saw fit. The input from the KWDP researchers was limited to providing some information about the growth characteristics of the new species; the trials were, in practice, almost totally farmer-controlled.

The project researchers then observed how the farmers managed their nurseries and where they planted their trees and why. This provided information about the farmers' knowledge of tree cultivation. Information on species performance was obtained through diameter and height measurements.

Opinions vary about the value of the results obtained from both sets of trials. The conditions in on-station trials are quite artificial and unrepresentative of what happens on farms. Many project staff feel they do not produce results which are directly relevant to extension. On-farm trials, on the other hand, do mirror real farming conditions. Another of their advantages is that they provoke a considerable amount of discussion between farmers, researchers, and extension workers and hence have a significant educational impact. Their biggest disadvantage is that they do not yield hard quantitative data.

Development of the technical packages has continued as the programme has proceeded. Perhaps the most successful exotic has been *Calliandra calothyrsus*. *Leucaena leucocephala* has had some success as a fodder species in warmer areas and *Sesbania sesban*, indigenous in Kenya, has been reintroduced in some places. *Mimosa scabrella* has shown somewhat unpredictable behaviour, but it often grows well in the colder parts of Kisii.

Making seeds rather than seedlings available

With the realisation that most planted trees come from seedlings grown by farmers themselves, the question was how best to support on-farm tree nurseries. Little or no relevant experience was available at the time.

Surveys and later contacts with farmers had made it clear that, though there is normally no lack of *Cupressus lusitanica* and eucalyptus seeds, the average farmer finds those of other species hard to obtain. This was particularly so in the case of some of the new, fast-growing exotics like *Calliandra calothyrsus* and was also true of the fuelwood species which hardly existed in many locations.

KWDP therefore developed the idea of a local "Seed Production Unit" (SPU). This consisted of an 0.05 hectare plot and was established by the project on communal land such as a school compound or outside the chief's office. The SPUs were fenced against livestock and planted with the fuelwood species which KWDP wished to promote, such as *Leucaena*, *Calliandra* and *Mimosa*. The idea was that the plots would produce seeds which would be freely available to any interested farmer. About 160 SPUs were established in the two project districts between 1984 and 1986.

Project staff found, however, that the SPUs did not fulfil their intended function. One technical reason was that many of the plots were too closely planted; some seed was produced round the edges but none from the inner trees.

More importantly, even when seed was produced, few people collected it.

This was mainly because they never saw the local SPU as belonging to them; it was a fenced woodlot "belonging to a project". Even SPUs on school compounds were overly respected, with teachers hesitating to collect the seed. In other cases, SPUs were established on farms and were more or less confiscated by the land owners — not surprisingly, since it was their own land.

The SPUs helped to demonstrate fuelwood species and their growth in the local communities, but they did not serve as effective local seed sources. Extension staff now believe that, as long as some farmers have the seed and are prepared to grow trees, the local community is more effective than the project at diffusing seeds to those who want them. KWDP now relies on free distribution or sale of seed packets to set the process in motion.

Improving the farm nurseries

KWDP agroforesters have tried to improve the traditional farm nursery model in a number of ways. The aim has been to increase the rate of survival in the nursery and after planting out.

One suggestion was that farmers should establish the nursery before the rainy season. Another was that the nursery bed should be raised so that seedlings could be lifted more easily, and without damaging the roots, when they were being transplanted. Other suggestions included pruning the roots with a panga (machete) and thinning the seedlings to produce healthy, vigorous growth — most of the farm nurseries are crowded with tiny seedlings.

The impact of the suggestions has varied. Not many farmers were prepared to carry out root pruning since they were not familiar with the practice and needed time to understand its relevance to the production of strong and healthy seedlings. Raising the seedbeds is, however, done more frequently and a considerable number of nurseries are now started in the dry season.

Extension methods

KWDP is not supposed to create its own extension service; its basic purpose is to develop appropriate techniques for promoting the planting of fuelwood trees. The intention is that these will then be taken over by the existing extension agencies in the Forest Department and the Ministry of Agriculture.

The number of KWDP extension workers is thus small. It consists of a team of 4-5 experienced and well trained officers in both Kakamega and Kisii. These are assisted by about 10 locally recruited "Village Energy Workers" in each district who liaise between farmers and the project.

In the initial phase of the programme, KWDP did not have a ready-made extension package. Its aim was to learn as much as possible about farmers' tree growing practices and attitudes before telling them what to do. Selected farmers were visited by an extension worker and were given five or ten seedlings to try. The idea was that the extension agents would learn from the farmers rather than just issuing orders and delivering a technical package.

Among the extension workers this was known as the "no message approach" and many had considerable difficulty in implementing it. They found that their frequent visits, carrying only 5-10 seedlings in a pick-up truck at a time, were incomprehensible to the farmers and somewhat embarrassing to themselves. Nevertheless, the experience undoubtedly had considerable educational value and taught the extension workers to appreciate and evaluate the knowledge possessed by farmers.

In order to reach a larger number of farmers, a group approach was then developed. This relied on existing farmers' groups or created new ones. These were visited by extension workers and encouraged to discuss fuelwood problems and take steps to deal with them. Among the activities encouraged were the creation of improved farm nurseries, planting woodlots for fuelwood, and inter-planting food crops with trees.

Working with such groups widened the impact of the programme while still leaving open the opportunity to receive effective feedback from farmers. It also provided valuable information on how the community perceived certain social issues. The fact, for example, that some questions emerged in discussions with individual farmers but were not discussed at a group level, indicated where a number of the more important social tensions and sensitivities lay.

At the same time, there were various practical problems which put a limit on the use of groups for extension. Because of funerals and other local events, group meetings were often cancelled. Local politicians and administrators tended to use the meetings for their own purposes. Moreover, it was found that the groups tended to be rather introverted and there was little spin-off to other local people.

The programme also attempted to use schools as part of the extension effort. Nurseries were established and run by pupils; demonstration plots were planted; and competitions and other activities were organised with assistance from KWDP.

One of the biggest problems found with using schools was the lack of effective communication between schoolchildren and their parents about the activities supported by the KWDP. Even when children planted fuelwood trees at home, parents often had no idea why they were doing so. Parents also had no sense of involvement with the fenced-off demonstration plots at school. Another problem in working with schools is that the activities are usually carried through only one teacher. If he leaves, it is the end of the school activity.

The project also undertook the task of raising the general level of awareness about the fuelwood issue among the whole population. The aim was to get husbands and wives to discuss it together. It was felt that this would help break down the reluctance of farmers to allocate resources to growing trees for fuelwood.

In Kakamega, this was done through the formation of an amateur drama group, assisted by professionals. The group performed a short play on the theme of the women's fuelwood shortage, confrontation between husband and wife, and the way out of the problems through planting trees for fuelwood. The drama was organised by KWDP and moved around in the district. The performances were held in schools, at markets, near churches and in other public places.

It was a popular activity which drew large audiences, but it had to be abandoned for organisational reasons. The actors became tired, or had to return to their normal jobs. Project staff also found themselves becoming over-burdened with the arrangements.

The project therefore decided to produce a film rather than relying on live performances. Since that time, films have been used and are found to be much more manageable than live drama. The film draws attention to fuelwood as a problem that needs to be discussed among men and women; it describes the establishment of farm nurseries for fuelwood trees; and it provides information on the management and harvesting of fuelwood trees.

Large numbers attended the film shows. In one area where the films were shown over a period of a month, almost 25,000 people, of whom about 70% were children, attended the showings. Most of the adults attending were given tree seeds and a follow-up survey of 840 of them found that 90% had planted the seeds in nurseries or directly.

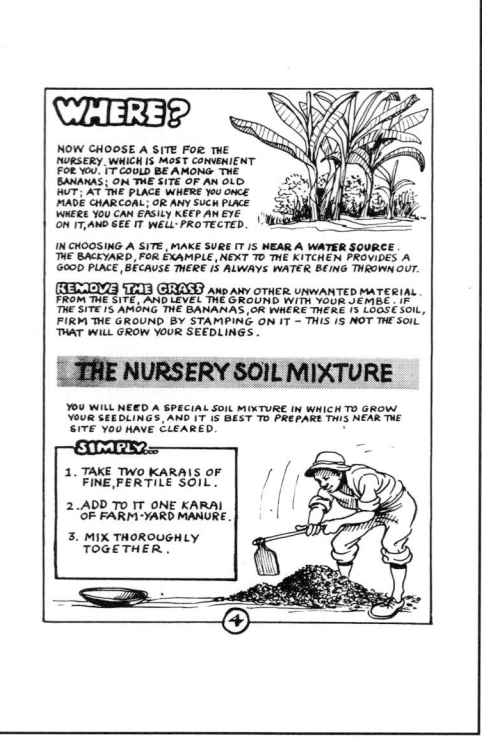

Figure 5.1. An illustration from a KWDP booklet.

Project staff in Kisii chose a different mass extension strategy. Public rallies were held in the different sub-locations and were addressed by senior local administrators and politicians. A sub-location has a population of about 10-15,000 people. So far, five rallies have been held and have each attracted attendances of 1,000-4,000 people. The project message was passed on through a drama group and was backed up by the various speakers. Pamphlets and packets of seed were distributed to promote a practical follow up.

The public rallies in Kisii have basically the same shortcoming as the drama in Kakamega: they require too many resources from project staff. While attendance is good, only a small fraction of the district population is reached in this way. There are over 100 sub-locations in Kisii and these could not be covered with the available project staff and resources.

There is, therefore, a clear need for cheaper and more effective extension methods. One approach presently being tested in Kisii is the use of radio messages in the local language. In conjunction with this, a system of selling packets of fuelwood tree seed is being established. The packets are sold by shopkeepers at a price of one Kenya shilling each (about 5 US cents). For the time being, this is heavily subsidised by the project, but it is intended that the system will eventually be self-supporting.

The KWDP pamphlets and booklets distributed in the two districts are well supplied with illustrations, such as those in Figure 5.1. Surveys indicate that the pamphlets serve a useful purpose. They are mostly written in English or KiSwahili, rather than vernacular, but this is done on purpose. Local people are used to speaking in vernacular at home, but most reading is done in English. However, many people are not used to reading at all and even the few sentences in project cartoons may be too much for them.

There is also a comprehensive farm nursery manual written by KWDP agroforesters. This is not suitable for the vast majority of farmers but is a useful reference for a motivated extension agent.

Monitoring the impact of the project

KWDP has invested considerable resources in measuring the impact of the programme. Each district has a monitoring unit consisting of one academically trained specialist assisted by a number of enumerators, and equipped with personal computers. The Nairobi headquarters formerly provided backup and looked after publications, but is now closed.

Enumerators are sent out with questionnaires to find out how many people attended a film show or a rally and whether the farmers have done anything with the information, leaflets and seeds they received. In a single survey as many as 600 people may be interviewed.

Various interesting results have been obtained in this way. For instance, the monitoring has demonstrated that the Kakamega films are having a significant impact. About 80% of the sample households had heard of the them, 68% attended, and 51% received some seed during the shows. In a sample of 800 households which obtained seed, 85% had established a tree nursery. It is not yet known how the seedlings raised in those nurseries are ultimately used (Njuguna, 1988).

In Kisii, a total of 911 households attending the rallies were surveyed. It was found that 810 households had obtained seed and, of these, 56% had created nurseries. A further survey covered 383 of the households which had established nurseries to find out what they did with the seedlings. This found

that 38% planted out seedlings; a further 21% had earlier done direct seeding; and 41% had not transplanted.

Some farmers said that pressure of work had prevented them from planting out but that they intended doing so later. Others said they had been unsure about when to plant out. This information was in the nursery booklet but had not been included in the nursery demonstration.

Mimosa scabrella growing among pineapples; the project views this tree as a fuelwood species but the farmer is planning to let it grow until it can be used for construction wood.

Almost half the seedlings were planted in cropland; of these, 69% were *Sesbania*, 15% were *Calliandra* and 12% were *Leucaena*. The other major planting site was in hedges which accounted for 30% of the total. Here, *Grevillea* accounted for 32% of the trees planted, black wattle 29% and *Calliandra* 19%. Most of the remaining planting was in woodlots in which 70% of the seedlings were black wattle.

Despite the data it produces, many extension agents feel that the monitoring has been academic and not very useful for extension. They say that it has served the need of the Nairobi headquarters to produce papers rather than being an instrument to adjust extension strategies. Perhaps this is a reflection of a degree of uncertainty about whether KWDP is primarily a research or an extension organisation.

Sustainability and replicability

Large resources have been available for KWDP, with well equipped project teams in Nairobi and in the two districts. The project has been active for about five years and several lessons can be drawn from the experience to date.

The importance of initial surveys in understanding local conditions is well illustrated. The existence of the farm nurseries, for example, had not been suspected until they were "discovered" by the survey. The survey also detected cultural constraints which had not been recognised before. It also established that, contrary to prevailing beliefs, the proportion of land under planted trees increased with population pressure.

The surveys have guided the project strategies and serve as an example to other projects. But in their current form, they are too time-consuming, complicated and expensive. This is especially so for aerial photo interpretation and data processing (KWDP, 1988).

Complicated district stratification is not necessary; it is normally sufficient to divide a district into 4-8 strata using existing data and common sense. This is illustrated by the fact that in Kakamega the agroforestry and cultural surveys were based on strata obtained from existing data. The remote sensing work was done afterwards, and has hardly influenced the subsequent project activities.

The agroforestry surveys have also been elaborate but this is understandable given the innovative character of the project. The farm nurseries of western Kenya are now widely quoted in forestry extension literature all over the world. But if the survey methodology is to be transferred to the government services, simplifications need to be made. These might, for example, include shortening the questionnaires to cover only the most relevant issues and reducing the sample size.

The extension methodologies offered by KWDP are not yet fully developed. The emphasis has been on creating mass awareness of fuelwood problems through films and rallies. These have been found effective in reaching large audiences. Moreover, most of the people attending them actually do something with the information and seeds they obtain. The problem is that organising such events is extremely expensive.

It is cheaper to use films but there are obvious difficulties. Running films in remote locations is, at a purely practical level, quite a difficult undertaking and places a heavy burden on project staff. Another problem is that of local relevance. The KWDP films are successful because they describe the particular situation in Kakamega. They are likely to be much less successful if used in other districts. It is unlikely that the ministries which take over the project will have the resources necessary to undertake an extensive programme of film making.

Another potential difficulty arises from the fact that the KWDP has operated more or less independently of Kenya Government institutions. This has given it considerable flexibility and allowed it to develop new and original methods of promoting tree growing. On the other hand, it means that the permanent institutions, like the Forest Department, have not had much involvement in the programme and are thus not yet in a position to carry it forward. The implication is that, if donor support is ended too early, the KWDP will have had little permanent impact.

The programme has, however, already generated a vast amount of data, knowledge and experience. It has shifted the emphasis from central tree nurseries to farm nurseries. It has also demonstrated the conflict of interests between men and women, and that the "fuelwood crisis" is a local social problem as much as anything else.

An important question about the basic relevance of the programme, nevertheless, remains unanswered. From the beginning, it has been based on the assumption that fuelwood shortage is a serious problem in women's lives and that, if cultural obstacles to growing fuelwood trees could be removed, the problem could be solved.

It is certainly true that a great deal of tree growing is taking place in the project area. But since no surveys have been carried out to assess the impact of the project, there is still little evidence that it has convinced farmers that growing trees specifically for fuelwood is an attractive proposition. Nor, indeed, is it certain that the majority of women see planting fuelwood trees as a serious way of easing fuelwood shortages, even after the awareness campaigns. There are cases, for example, of women who planted hundreds of fuelwood trees with assistance from KWDP, but who have lost interest and replaced the *Leucaena* and *Calliandra calothyrsus* with eucalyptus.

` The project is increasingly recognising that both men and women want to plant trees for a variety of reasons, not simply for fuelwood. One extension agent has complained that the KWDP package should have been broadened to include eucalyptus, *Grevillea* and other timber trees. It is significant that the project has recently changed its name from the "Kenya Woodfuel Development Programme" to the "Kenya Woodfuel and Agroforestry Project".

References:

BRADLEY, P.N. (1988). "Survey of woody biomass on farms in western Kenya". Ambio

Vol.17, No.1.

CHAVANGI, N. (1984). "Cultural aspects of the fuelwood programme". KWDP working paper No.4.

VAN GELDER, A. and P. KERKHOF (1984). "The agroforestry survey in Kakamega District". KWDP working paper No.3.

KOLA, H.O. (1987). "The outcome of mass rally. An overview of on-spot monitoring". Kisii Technical Report No. 1, KWDP.

KUYPER, J. (1988). "On farm agroforestry trials in Kisii District". KWDP working paper No.12.

KUYPER, J. and P. BRADLEY (1985). "Woodfuel and agroforestry in Kisii District". KWDP working paper No.7.

KWDP (1988). "A critical analysis of KWDP activities Kisii District and work programme up to 1991". KWDP.

MBUTHIA, M. (1987). "Spacing trials: first harvest data analysis report". KWDP (draft).

MUNG'ALA, P.M., J. KUYPER and S. KIMWE (1988). "On-farm tree nurseries". KWDP.

NJUGUNA, M.N. (1988). "Kakamega film monitoring - Phase Two". KWDP (draft).

ONG'AYO, M. (1986). "Report on the cultural survey carried out in Kisii District." KWDP.

BAT AFFORESTATION SCHEME, Kenya

TREE GROWING, BUT NOT FOR FUELWOOD

Farmers who cultivate tobacco in Kenya are under a contractual obligation to grow trees. These are supposed to provide the fuel for curing the tobacco.

The scheme has been a success in that farmers have grown large numbers of trees. But they prefer to use these for poles or construction timber while continuing to cut the natural woodlands for fuel.

Tobacco growing in Kenya

With the collapse of the East African Community during the 1970s, Kenya found itself completely dependent on imports for its tobacco supplies. A start was therefore made on indigenous tobacco growing. By 1982, the country was meeting its full needs and production was running at 4,650 tonnes of cured tobacco. Output has continued to rise and was 8,600 tonnes in 1986.

Tobacco production is managed by the multinational British American Tobacco Company Ltd (BAT), which is also a major tobacco producer in many other African countries. It uses a system in which individual farmers, called out-growers, grow the tobacco and sell it to the company. The farmers are supplied with fertiliser, tobacco seedlings and pesticides on credit during the growing season. They are also given the necessary training through the BAT extension service. When the crop is sold to the local leaf collection centre of BAT, the money owed by the farmers is deducted from the amount paid.

The farmers cure the tobacco before selling it. The curing takes place in a large shed, called a barn, which the farmer himself builds; some farmers build several. As they are thatched wooden structures, they are prone to fire and sometimes burn down.

Tobacco curing is a forced drying process, in which the freshly picked leaves are exposed to hot air. Depending on the type of tobacco leaf, this is done either by flue-curing or fire-curing. Flue curing uses a system of metal pipes which

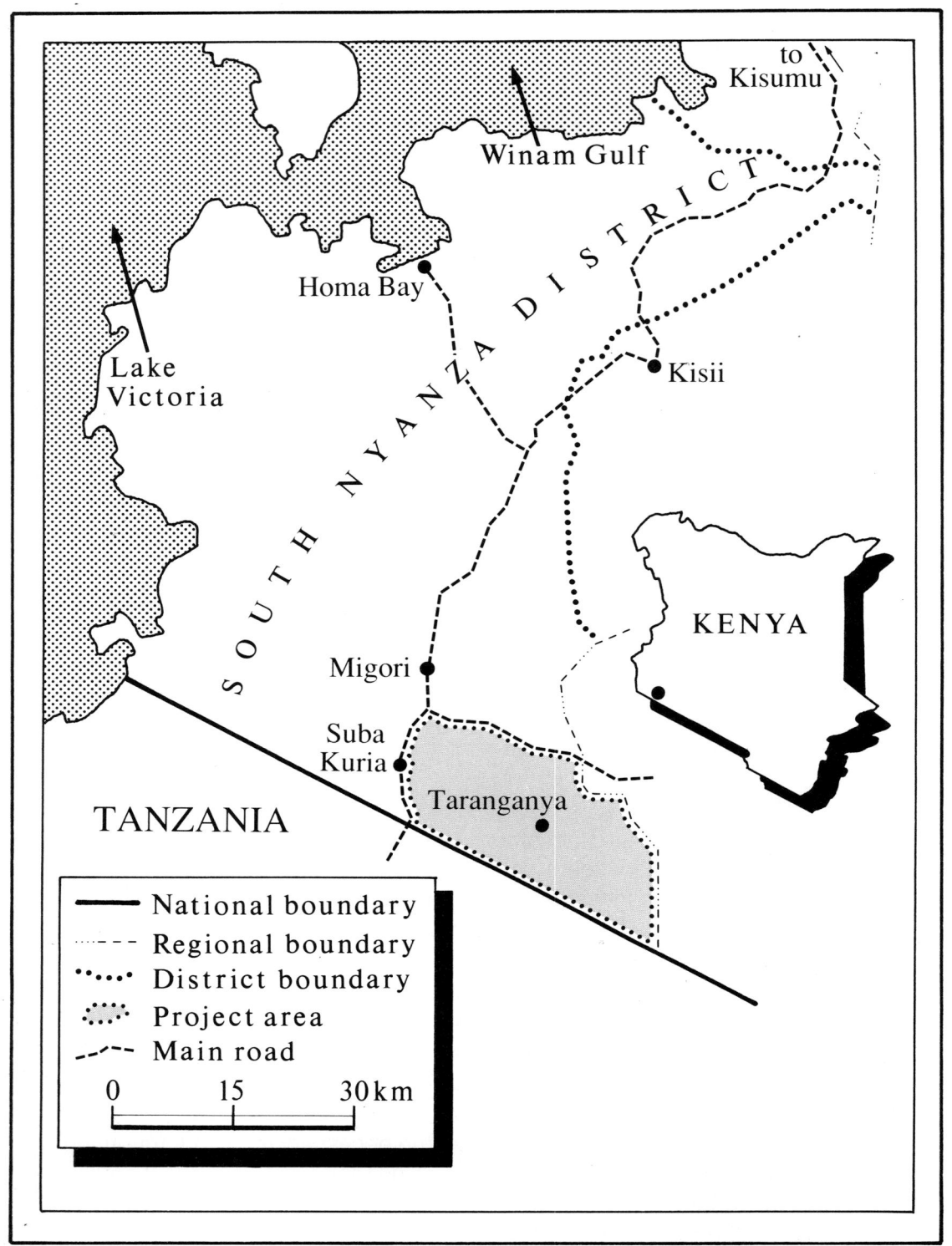

Lake Victoria

Winam Gulf

to Kisumu

Homa Bay

Kisii

S O U T H   N Y A N Z A   D I S T R I C T

KENYA

Migori

Suba Kuria

Taranganya

TANZANIA

National boundary
Regional boundary
District boundary
Project area
Main road

| 0 | 15 | 30 km |

Name of project:	BAT Afforestation Scheme
Address:	BAT, P.O.Box 2, Kehancha, Kenya
Project area:	Kuria Division, West Kenya
Average rainfall:	1,500 mm per year
Implementation:	British American Tobacco Company
Funding:	British American Tobacco Company
Exchange Rate:	18 KSh = 1 US$ (early 1989)

heat the air inside the barn by conduction, whereas fire curing dries the tobacco directly with the hot gases from the fire.

The amount of wood required for flue curing varies greatly depending on the design of the barn, the amount of tobacco packed into it and the moisture content of the wood. Figures of up to 40kg of wood per kilogram of tobacco have been found but more normally the quantity lies in the range 4-20kg per kilogram of tobacco. The amount of wood used in fire-curing is considerably less but the price paid for the tobacco by the company is also lower.

Most of the tobacco is grown in the west and central parts of the country with the main production taking place around Kuria in western Kenya. Kuria is part of the Kenya Highlands and has an undulating landscape with an average altitude of about 1,500 metres. The average annual rainfall is 1,500mm and there are two rainy seasons in the year. Maize, cassava, sweet potato, beans and banana are the major food crops and tobacco and coffee are the predominant cash crops.

The local people used to be pastoralists but are now settled farmers. They still maintain substantial numbers of cattle which they keep at night in a yard to protect them against the attacks by Masai herders and other outsiders. The protected yard is formed by building the houses of an extended family in a wide circle and connecting them with a wall of stakes. The population density in the area is high and increasing rapidly, reaching over 200 persons per square kilometre in some parts.

Women stacking firewood next to a traditional barn for flue curing tobacco.

Kuria was selected by BAT during the 1970s partly because of its suitability for tobacco growing but also because of the abundance of its natural forest and bush cover. Farmers who wished to become out-growers had to have a large landholding with a stock of scrub and trees sufficient to meet their fuel needs for at least seven years. In addition, they were restricted to a maximum area of 0.5 hectares of tobacco.

65

Reducing the environmental impact of tobacco-growing

In order to reduce the cutting of natural woodlands for tobacco curing, BAT has introduced a number of technical measures to improve the energy efficiency of the barns. As a result, the average wood consumption has been reduced to 8 kg per kilogram for flue-cured, and about 2 kg per kilogram of tobacco for fire-cured. In Kuria, the company has also shifted the balance away from flue-curing to the less energy intensive fire-curing (IFSC, 1986).

In addition, BAT has developed a scheme under which all farmers growing tobacco for the company must plant a substantial number of trees each year to provide for their future fuel needs. Those producing flue-cured tobacco must plant 1,000 seedlings per year; those producing fire-cured tobacco must plant 100. In some years, the required numbers have been increased to 1,500 for flue cured, and 300 for fire cured. Tobacco farmers must plant trees for at least three years.

Initially, the species planted were almost entirely eucalyptus with a few *Markhaemia platycalyx* and *Cassia siamea*. The recommended spacing was 2x2 metres but many farmers planted closer, perhaps as a response to the shortage of land in the area. By 1987 a total of 13.5 million trees, or about 3,600 hectares, had been planted, most in Kuria and its surroundings.

The seedlings used to be produced at nurseries at the leaf collection centres, with some of the larger nurseries producing up to a million seedlings annually. Distribution from these central nurseries, however, turned out to be expensive and inefficient. Roads are often impassable in the wet season and in any case it was impossible to deliver the seedlings to the farm gate of thousands of individual farmers. As a result, seedlings were often left at the roadside for farmers to pick up, which led to considerable losses.

A study in 1983 found that the overall seedling survival rate was only 30-50%, a low figure for the moist mountain climate of Kuria. BAT was

BAT runs 11 tree nurseries throughout the tobacco growing areas, but has encountered problems with seedling distribution.

therefore forced to look for an alternative approach. The method now in use relies on local nurseries which are part of the tobacco production cycle.

Tobacco must be raised in a nursery before it can be planted out at the beginning of the rainy season. This is done in local nurseries, organised by a BAT extension worker, in which each farmer has his own tobacco seedling bed. Farmers are now required to raise tree seedlings alongside their tobacco plants. This has solved the distribution problem and reduced the cost of seedling production for the company.

Obviously, many of the farmers would prefer not to plant trees. Since the farmers meet regularly under the guidance of the BAT extension worker, the local nursery system provides ample opportunities for reminding the farmers of their obligations. Those who do not look after their trees properly find that some of the necessary inputs for their tobacco growing are withheld.

The ability of BAT to lay down the rules in this manner depends upon the fact that the farmers do not have any other sales outlet for their tobacco and that the price they are being paid makes it more attractive than other crops, even taking into account the land and labour required for the tree growing. The company has attempted to persuade tobacco growers to grow trees for fuel in a number of other countries but with less success than in Kenya.

Choice of tree species

The heavy emphasis on eucalyptus has led to criticism by Kenyan politicians and leaders who, for a variety of reasons, have been showing an increasing antipathy to the species. BAT has responded to such pressures by promoting indigenous species but these are not popular with the majority of farmers.

The farmers are encouraged to collect their own seeds and mainly choose eucalyptus and *Cupressus lusitanica*. Extension workers may provide seed from indigenous species but most farmers are not prepared to raise a significant

Paul Kerkhof/Panos Pictures

number of these. They feel that, if they have to plant trees, they want them to be straight and fast growing.

Researchers at Moi University have estimated the production of eucalyptus in the tobacco zone at 20 cubic metres per year for the first cycle and 24 cubic metres for coppice rotations. Their estimate of the yield from fast growing indigenous species in the same area is only

A section of this tobacco nursery is devoted to tree seedlings; farmers now have to look after their own trees.

5-10 cubic metres. The reluctance of the farmers to change to the indigenous species is therefore understandable.

In 1982, senior staff of the company held a seminar with ICRAF on afforestation strategies. This created a certain amount of enthusiasm for promoting intercropping with leguminous species as an alternative to the eucalyptus woodlots. *Leucaena leucocephala* and other nitrogen fixing trees were raised in the central nurseries and were distributed to the farmers. But the farmers apparently had little success and soon refused to plant any more of these leguminous exotics.

'We don't grow trees to burn them'

The afforestation scheme has now been active for over a decade in Kuria and millions of trees have been successfully grown. But contrary to the objective of the scheme, the trees are hardly used for curing tobacco. This has been observed not only in Kuria but also in the other tobacco growing areas. One survey in Kuria found that only 5% of the tobacco farmers used eucalyptus as the principal fuel for curing while the rest used trees and shrubs from the natural woodlands as well as agricultural residues.

Farmers say that they do not want to grow good quality construction wood and then burn it. As long as there is wood to be obtained from the natural woodlands, it will be used as fuel for curing instead of the highly valued trees growing in the woodlots. The net result of the tobacco scheme is thus a rapid substitution of the indigenous trees by eucalyptus and cypress (Kerkhof, 1987).

This does not mean, however, that the scheme is without value. This is clearly illustrated by the case of Mwita Kihika, a farmer, who has been a registered tobacco grower since 1976. In the beginning, he planted 1,000-1,500 trees a year in accordance with his contract. As these were maturing, he used the indigenous trees on his lands for fuel. When these trees were all finished he bought wood from his neighbours.

By now, his woodlots have increased to 10,000 eucalyptus trees, some of which he has started selling for construction wood. But he is also finding that it has become very difficult to obtain fuel supplies from natural woodlands. He has therefore begun to use his eucalyptus trees for tobacco curing.

It thus seems clear that as long as farmers are able to obtain woodfuel from the natural woodlands, whether for tobacco curing or any other purpose, they will continue to do so. Planting woodlots will not prevent this happening. But, in the meantime, planting woodlots does provide farmers with a cash income; and, in the longer term, it ensures that there are still wood resources available to provide fuel for the tobacco industry.

References:

IFSC (1986). "The use of wood by the tobacco industry in Kenya". International Forest Science Consultants.

KERKHOF, P. (1987). "South Nyanza District Afforestation Programme". Kenya Forest Department/Danida.

RURAL AFFORESTATION PROJECT, Zimbabwe

EVOLVING AFFORESTATION STRATEGIES

Zimbabwe's communal lands, the home of the majority of the country's rural people, were badly neglected up to the time of independence in 1980. Since then, there has been considerable government concern about the deforestation and land degradation taking place in these areas.

A major programme to promote the production of fuelwood and poles by farmers and boost the rate of reforestation was launched in 1983. Experience has shown the need to decentralise seedling production and increase the involvement of local communities in planning and decision-making. The initial results from an NGO programme in the Zvishavane area support this view. As a result, the Forestry Commission is now making major changes in its strategies.

The Rural Afforestation Project

A national study of rural wood consumption was carried out by the Whitsun Foundation of Zimbabwe in 1981. One of its conclusions was that the country's woodlands were being rapidly depleted. Cutting for fuelwood was diagnosed as the major cause and measures to deal with the problem were proposed.

In 1982, the Forestry Commission prepared a proposal for a US$28 million Rural Afforestation Project. The main objective was to increase fuelwood and pole production by small farmers in the communal lands. These cover about 40% of the country and are held, generally under traditional land tenure systems, by African as opposed to white farmers.

An agreement for a soft loan of US$10.8 million was signed with the World Bank in 1983. This originally covered a period of four years but has now been extended to 1989.

The project began by creating the necessary institutional and infrastructural basis for its activities. A Rural Afforestation Division was established within the Forestry Commission. The creation and staffing of this division absorbed a

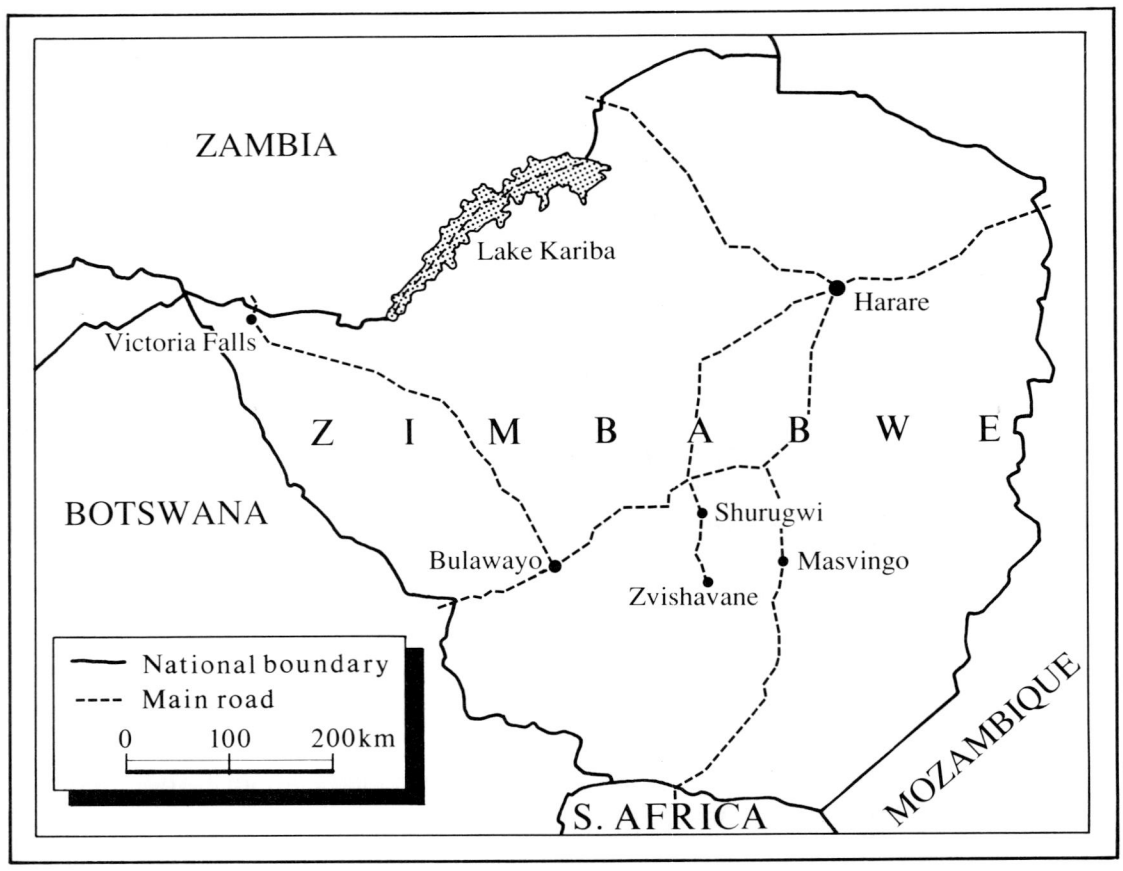

Name of project:	**Rural Afforestation Project**
Address:	**Forestry Commission, PO Box 8111, Causeway, Harare. Tel: 736216**
	ENDA, PO Box 3492, Harare. Tel: 708568
Project area:	**Zimbabwe's communal lands**
Average rainfall:	**Average for Zimbabwe 500-1,000 mm per year**
Implementation:	**Forestry Commission**
Funding:	**1983/88 - World Bank: US$ 7.3 million**
	Government of Zimbabwe: US$ 3.5 million
Exchange rate:	**2.3 Zimbabwe $ = 1 US$ (October 1989)**

considerable proportion of the initial project funding; but it has provided the Forestry Commission, for the first time in its history, with a section assigned to deal, on a national basis, with tree planting by rural people.

Nurseries

The production and distribution of seedlings is the principal field component of the project and has gradually been expanding. By 1988, some 70 nurseries had been established and the total annual production was about 3 million seedlings. The vast majority of these are eucalyptus which is well known to

Chris Pennarts/Panos Pictures

Although there is pressure on wood resources in Zimbabwe's communal lands, surveys have shown that wood shortages are not yet seen as a serious problem by local people.

Zimbabwean foresters as straight and fast growing. The seedlings are sold at a subsidised price but the take-up by farmers has been very poor and this component of the project is in considerable difficulty. Only half the nurseries have managed to sell more than 50% of their seedlings. One estimate of their impact is that, up to the end of 1988, they had served well under 1% of the rural population.

Demonstration woodlots, and fuelwood plantations

Demonstration woodlots were established at the nursery sites to illustrate proper soil preparation, application of chemical fertilisers, the use of pesticides against termites, and other sylvicultural operations carried out in conventional forestry. Although the impact of these woodlots has not been formally monitored, it is unlikely that many farmers have been prepared to follow the procedures demonstrated.

Principally, this is because the concept of a woodlot is alien to most farmers. Moreover, the material and labour inputs required are likely to be outside the reach of most farmers in the communal lands. A total of about 35a of demonstration woodlots were planted but this project component has now been more or less abandoned.

The project also envisaged the creation of some 1,400 hectares of block plantations to meet rural and urban fuelwood needs. This component has also been abandoned as it was found that the plantations were not cost effective.

Chris Pennarts/Panos Pictures

Extension efforts in schools have been one of the most successful parts of the programme.

Publicity campaigns and extension services

The project has also carried out a publicity campaign to raise public awareness of the deterioration of the country's woodlands and the need for tree planting. Large quantities of posters, calendars and other materials have been distributed and there have also been radio and television programmes.

Competitions in which 75% of the country's schools participated have been held to stimulate tree planting. A fund has been created to provide assistance in kind, especially fencing and pots, to school and other groups and to individuals wishing to establish nurseries. Although monitoring data are scarce, many project staff feel that this aspect of the project has been relatively successful.

An agroforestry training officer was appointed to Agritex, the government's agricultural extension service, to train agricultural extension workers in agroforestry. Some successful promotional activities have been reported, such as the establishment of school nurseries, but the overall role of Agritex in the project to date has been limited.

Some foresters question whether Agritex has fully accepted that trees have an important role in farmland. Agritex, in its turn, has pointed out that the foresters have not been able to present credible agroforestry packages and that there is therefore not a great deal they can tell farmers about agroforestry.

Both sides have a point. It is clear that simply adding an agroforestry adviser to the agricultural extension service does not turn it into an effective agroforestry extension organisation. It is also obvious that there is a need for proven agroforestry packages which can be made available to farmers in the different climatic zones of the country.

Major baseline survey

A major baseline survey of small farmers' attitudes and approaches to tree growing was carried out by consultants in 1985. A total of 1,829 farmers were surveyed using an elaborate pre-coded questionnaire with the data being processed by computer. Aerial photographic surveys were also carried out.

The survey found that deforestation is strongly correlated with clearing land for cultivation. Non-arable land has not been deforested and may even have more woody biomass than 20 years ago. Deforestation is thus not simply a matter of overcutting for fuelwood and poles by rural families. The deforestation problem, if it is a problem, is therefore not going to be solved by woodlots.

Another conclusion was that fuelwood scarcities are not as serious as assumed in the project proposal. Most rural people apparently find the provision of fuelwood and poles a relatively minor problem.

The survey also revealed that about 60% of the questionnaire respondents had planted trees in the previous five years. Fruit trees are the most popular, followed by trees for shade or ornament, and then for poles. Planting of trees for fuelwood, the initial focus of the project, is normally not considered by rural people. Where, eucalyptus is planted it is for poles or timber and not for a fuel.

These findings clearly showed that the project strategy of promoting eucalyptus woodlots for fuel required drastic changes. The fact that the survey was carried out separately from the Forestry Commission and Ministry of Agriculture meant, however, that it did not incorporate the ideas of the extension staff; neither have its results been well communicated to them. The survey report is also rather academic, which further restricts the access of project staff to its results.

The conclusions clearly illustrate the risk in delaying a baseline survey until after project commencement. In this case, it showed that the small farmers' interest in trees extended far beyond eucalyptus woodlots. The narrow initial focus of the project could have been avoided had the survey been carried out in good time.

Research, monitoring and evaluation

In 1982, a research programme, called the Dry Zone Afforestation Project, was set up with funding from IDRC, of Canada. The initial focus was upon on-station screening and provenance trials of eucalyptus species for dry zones, in which 76% of the communal areas fall. In recent years, the project has included trials of indigenous acacias and *Azadirachta indica*. The programme complements the work of the Forestry Commission's field stations in the Eastern Highlands which have a long history of commercial forestry research.

In 1986, a species trial project, assisted by an Australian grant and focusing primarily upon Australian species, was initiated. This uses traditional forestry methodology and is mainly concerned with screening eucalyptus and *acacia* species.

A Monitoring and Evaluation unit was in operation during the first phase of the

Rural Afforestation Programme. It was not, however, particularly effective. Seedling survival surveys, for example, were carried out using questionnaires. These were sent to all district nursery managers, who were asked to count seedling survival in their districts.

The sampling strategy was left to the nursery managers and many did not respond. The potential for bias in data collected in such a manner is so great that no reliance can be placed on the results obtained. The Rural Afforestation Division is, however, fully aware of the deficiencies in its monitoring and evaluation and major changes are currently being made.

Emerging new strategies

The first phase of the Rural Afforestation Project was characterised by the production of eucalyptus seedlings in centralised nurseries. According to some foresters, this traditionalist approach was virtually inevitable. The project started in the early 1980s, just after independence, when Zimbabwean foresters had little

Fruit trees have proved to be much more popular than eucalyptus, especially amongst women.

experience beyond conventional plantation forestry.

There are several obvious reasons for the disappointing results obtained. One is that farmers had to travel long distances to obtain seedlings. When the project was being designed, it was assumed they would be willing to do this. But it turned out not to be the case in practice. Another reason for the low response was that farmers were interested in other species besides eucalyptus. Fruit trees and a variety of indigenous species, for instance, are often popular but were not provided at the nurseries.

A variety of new strategies are now emerging. These are shifting the emphasis of the project to the support of farmer and community initiatives. Seedling production is being directed away from large central nurseries to a much greater number of farm and community nurseries. The choice of seedlings is being widened to include fruit trees and indigenous species.

At a technical level, the focus is on cultivation methods which are adapted to the capacities of the small farmer. This means abandoning many of the conventional guidelines for tree planting, such as the application of commercial fertilisers and pesticides.

Such changes do not come easily for some forestry staff. Senior project staff, as well as diploma holders such as district nursery managers, have to reorient their attitudes almost completely. The "eucalyptus woodlot model" is still firmly established in the minds of many.

In one district, the forester in charge of the nursery insisted, "Farmers are not interested in fruit trees. That's why we have only eucalyptus here." Yet the owner of a farm nursery supported by the same forester bitterly complained that he wanted to raise fruit trees but could only obtain eucalyptus seed. But, he said, "I will continue to pester the forester until he gives me the seeds I really want."

The project is also exploring the potential for improved management of the indigenous woodlands by local communities. To this end, a research project was started in one district in 1988 to explore how to employ "Diagnosis and Design" techniques in collaboration with the local population. The intention is to use village meetings to analyse the problems facing the community and develop methods for managing the local woodlands and producing the range of seedlings the community requires.

It will take some time before the full impact of these new strategies is felt. Nevertheless, the change of emphasis is already evident. Decentralisation of seedling production, for example, is well under way and increasing numbers of farm, school, and village nurseries are being assisted by the project.

To provide support for the new approach, a multi-disciplinary planning unit, employing an economist, a sociologist and an ecologist, has been established in the Forestry Commission. The unit is also charged with monitoring and evaluation.

Changes are also required in the curriculum at the Zimbabwe College of Forestry if the foresters of the future are to be equipped with the necessary skills and attitudes. Such reforms cannot be brought about overnight and it will be some time before their effects are reflected in the attitudes of new graduates. Nevertheless, it is clear that the Forestry Commission is in the process of creating a new and profoundly different attitude among foresters to the promotion of tree growing by rural people.

The Zvishavane Project:

A small tree growing and woodland management programme in the Zvishavane area provides an interesting contrast with the centralised approach taken by the Rural Afforestation Project. The programme is promoted by ENDA, an environmental NGO, and relies upon a gradual build-up of community confidence and activity. This is a necessarily slow process with little immediate impact. The question is whether it provides a model which can be generalised or replicated at a national level.

An alternative approach

The Zvishavane area is in the communal lands close to the famous Great Zimbabwe ruins. The climate is relatively dry with an annual rainfall of 450-650mm per year. The land is slightly undulating, with villages and farmland scattered in dry woodlands. There has been a major reduction in the woodland cover in the area during the past few decades.

The project had its origins in a study carried out by two PhD students in 1985-87 which examined the way the indigenous woodlands in the area were managed by local communities. This revealed that there was a considerable depth of knowledge about trees among the farmers in the area; it also showed the importance of the woodlands to the community. A further finding was that people believed deforestation was occurring because the government and outsiders wanted to get rid of the indigenous woodlands.

The study also made it clear that the local people were concerned not just with the woodlands as such, but with their structure and composition. They were well aware of how the woodlands could be managed by planting and cutting in accordance with local rules. A basis for improved local management systems was thus in place, though it clearly needed support and improvement.

ENDA was interested in the study, not only because it contributed to an increased understanding of Zimbabwe's natural woodlands, but because it also offered an opportunity to develop a community based project. A project proposal was drawn up and was accepted by the Ford Foundation. The first phase of the programme is for a period of four years with funding of US$80,000 per year.

In 1987, the field assistant in the PhD study was taken on as the first community worker and in 1988 a further three community workers and a local co-ordinator were employed.

The difficult path from consultation to action

The aim of the project is to improve the local management of woodland resources. It is based on the assumption that its efforts will be more effective if they are based on existing knowledge and practices and take into account local needs and priorities.

Initially, the community worker interviews 30-40 people in a village in order to document their perceptions of their woodland resources. This takes about a week. The community worker then calls a village meeting to discuss his findings and together with the villagers draws up a woodland resource management plan to meet their particular needs. During 1987-88, a total of 25 village meetings were held at which resource management plans were agreed. Most of these called for planting in grazing areas, around homes and to a lesser extent on crop lands. The total number of seedlings requested was 35,000.

The survey method is extremely simple compared with that of the Rural Afforestation Project. Although it yields very few quantitative data, its advantage is that the local staff and farmers understand what it is for and what its results mean. The findings to date make it abundantly clear that farmers are aware that the woodlands are being depleted, and are interested in action to halt the process. They want to plant trees, especially fruit trees. Schools are also interested in tree planting.

The project has established four tree nurseries. Although there was little information available on the survival and management of the indigenous trees

A nursery run by a private farmer; the project provides plastic bags and seed.

which were planted, a total of 5,000 seedlings were produced in the first year. These were taken to the planting sites using a variety of means of transport, ranging from ox-carts to buses; heavy use was also made of the ENDA vehicle.

A total of seven villages have also created their own fenced woodlot of indigenous trees. These are meant to increase local awareness of tree growing and act as a focus for community activity. In principle, the tree species are selected by the villages themselves but the community worker has tended to favour indigenous trees. Fencing is provided by the project; without it, seedling survival is virtually nil.

These village woodlots may well fulfil their purpose of raising community awareness about indigenous trees but the balance between "community initiative" and "project priority" is a precarious one. Many farmers are primarily interested in fruit trees, such as the papaya. It is not clear whether many villagers are really interested in establishing woodlots of indigenous species.

In discussions on possible community rules and restrictions which might be used to reduce the damage being done to the local woodlands, the following are commonly suggested:

● Do not fell trees by burning as this kills the tree and prevents coppicing

● Fell trees in a dispersed way rather than clearing clusters of trees

● Do not cut heavily browsed species below a height of 1-2 metres

● Cut branches rather than the whole tree; if a whole tree is cut the whole tree should be used

● Do not cut whole trees simply to get leaves as dry season feed.

There is nothing new about such regulations; they were in operation during the colonial period. The question is whether they can be accepted and adhered to by the present communities. The development of such systems of self-regulation is an important objective of the programme but it is too early to say whether tangible results will be obtained.

Lessons from the project

This project was developed partially as a reaction against the approach initially taken by the Forestry Commission and Agritex. The background information acquired during the 1985-87 study provided a starting point for dialogue with the community. As a result of this and the consultation carried by the project staff, it has been possible to tailor extension activities to the priorities of the local people.

The project is, however, still at a very early stage and faces a variety of constraints. The problems of effective seedling production and distribution, for example, are similar to those experienced in tree growing programmes elsewhere. Unless an effective system of decentralised seedling production can be developed, the present heavy reliance on distribution by motor vehicle will continue. But the consultation which is being carried out is, at least, able to ensure that the seedlings which are produced are those preferred by the community.

The problem of how to promote improved management of the natural woodlands by the local community remains unsolved. While there is a general measure of agreement about what needs to be done, the community discipline and social control required to ensure adherence to the necessary rules and regulations remain to be developed. Such changes are slow to take place and if they are to happen, the project will require the four years planned for Phase I and probably longer.

There are also questions over some of the programme activities. At this stage, it is difficult to see the point in tiny fenced-off plots of slow-growing indigenous trees. More time will be required before it is possible to assess the value and direct, or indirect, impact of this component of the programme.

At times, there can also be a difference between what the local community wants and what the project is suggesting or offering. One village, for example, wanted to plant a eucalyptus woodlot rather than one of indigenous species. Although eucalyptus woodlots are not particularly favoured by the project, the extension worker had to concede to the wishes of the village.

A balance obviously has to be struck between the concept of full community control of the programme and the objectives and preferences of the project staff. Such contradictions are already difficult to resolve and would be a great deal more acute if the project were to be extended throughout the country.

There is thus a great deal more to be learned during the remaining years of Phase I of the programme. Nevertheless, it is already clear that there are tangible benefits to be obtained from a careful study of local attitudes and tree growing practices before launching programmes to promote tree growing. It is equally clear that it is necessary to ensure that there is regular consultation with local people as the programme is developed and implemented.

References:

GUMBO, D. J. (1989). "Community-based management of indigenous woodlands: the Chivi and Zvishavane Districts Demonstration project". ENDA, Harare.

WHITSUN FOUNDATION (1981). "Rural afforestation study".

WORLD BANK/IDA (1983). "Zimbabwe: Rural Afforestation Project – Staff appraisal report."

GURSUM LAND USE PROJECT, Ethiopia

PROTECTING THE SOIL IN ETHIOPIA'S HIGHLANDS

The land in Gursum is poor and over-grazed. Erosion has stripped away much of the soil from the mountain slopes and rangelands. Deforestation has been aggravated by the government's policy of forced villagisation.

A major effort to improve and protect the land by getting farmers to grow trees has been under way for the past three years. Bureaucratic rigidities, over-dependence on food aid, and the security problems of the area are making progress difficult. But, given the obstacles, the project has made considerable progress.

The project area

The Gursum Land Use Project (GLUP) is situated in the Hararghe Region of the Eastern Highlands on the border with Somalia. The region covers an area of some 25,000 square kilometres and has a population of about 4 million people. Most live in the higher altitude areas, at a height of 1,800 metres or more above sea level.

Only the mountain plateaux, which occupy a tiny proportion of the area, are suitable for crop cultivation. They receive an annual rainfall of 1,000mm and are densely populated. The rest of the area is dry scrubland fit only for poor grazing.

Like most of the country, Hararghe has been through a great deal of turmoil in the past decades. The war with Somalia has several times turned the area into a battlefield. Remnants of tanks and other military equipment are still scattered around parts of the region. Internal unrest, officially referred to as the activity of "bandits", is currently a major security problem. Severe famine has also hit the area several times.

Villagisation was enforced in 1983 in accordance with the policy of the revolutionary government which came to power in 1974. Paintings of Marx,

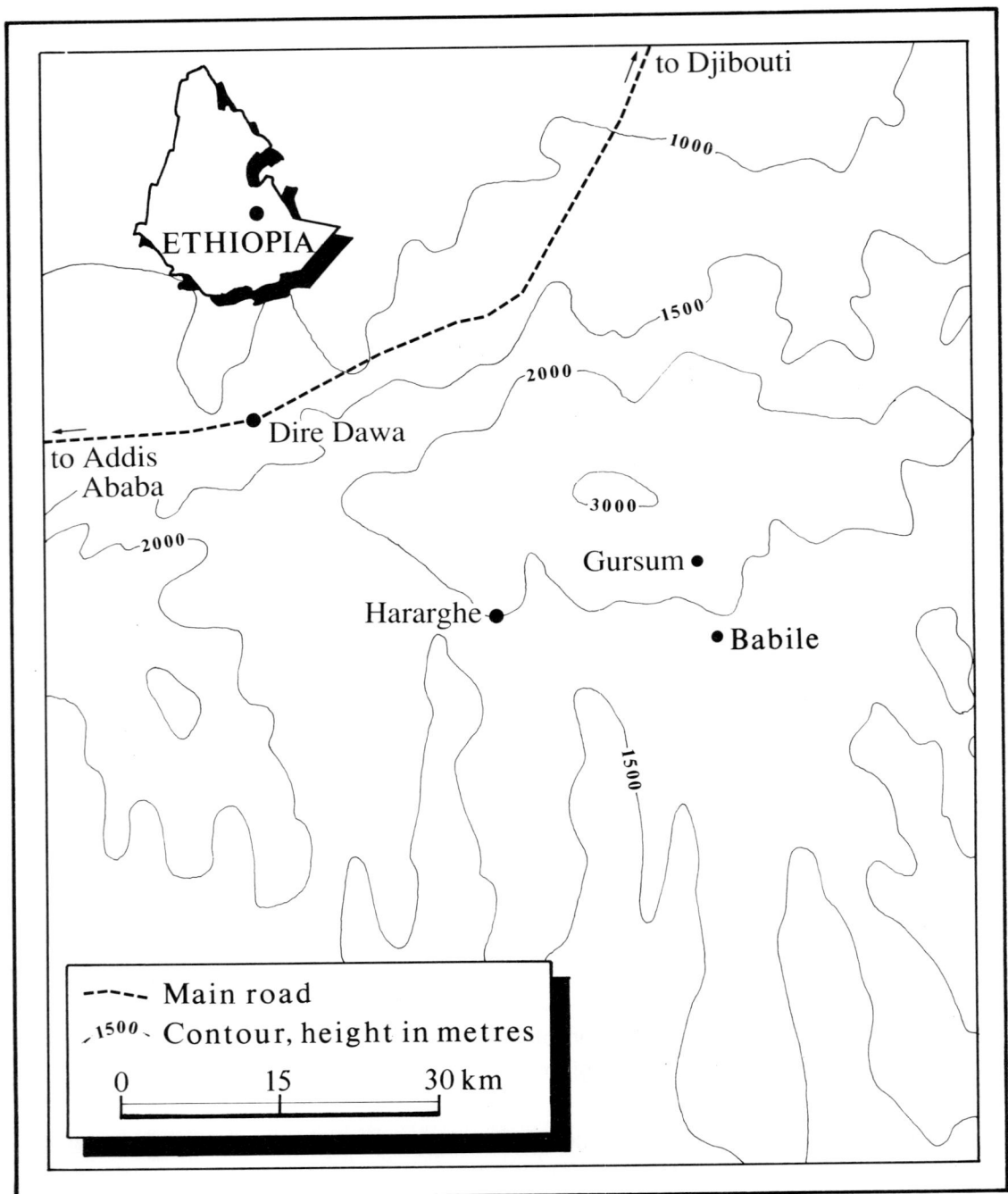

Name of project:	**Gursum Land Use Project (GLUP)**
Address:	**P.O. Box 454, Dera Dawa, Ethiopia. Tel: 111233**
Project area:	**Gursum and Babile Districts, Hararghe Region**
Average rainfall:	**1,000 mm per year (mountain areas)**
Implementation:	**CARE (Ethiopia) and Ministry of Agriculture**
Funding:	**CARE/Band Aid: $250,000 per year**
	USAID: Several thousand tonnes food per year
Exchange Rate:	**2 Birr = 1 $US (October 1989)**

Lenin and Engels dominate the town and village markets. Only a small number of isolated pockets of population high up in the mountains remained outside the scope of the villagisation drive.

The new villages consist of rows of neatly arranged houses, each one surrounded by a tiny garden. Families also have their own land holdings which vary in size between 0.5 and 5 hectares. This land is owned individually and inherited from father to son, but cannot be sold. With the growing pressure on land resources, young people and newcomers receive an increasingly smaller piece of land. Land transfers are arranged through the Peasant Associations (PAs) which are the governing bodies for clusters of villages.

The village communities grow sorghum, sweet potatoes and a small grain called tef, but food aid still remains important. Hills, river valleys and dry bush land are used for grazing cattle, sheep and goats. Many parts are over-grazed and show signs of soil erosion and land degradation. Much of the area has been deforested for ages, but concentrating people in villages has put further pressure on the remaining tree resources.

Producing seedlings and getting them planted

The project began in 1986 and is implemented by CARE in collaboration with the Ministry of Agriculture. The area covered consists of the Awraja (districts) of Gursum and Babile. Funding is partly in cash and partly in food and has been provided mainly by CARE's own funds, Band Aid and USAID. The yearly expenditure in 1987 and 1988 was about US$250,000 together with several thousand tonnes of food aid.

The main objectives of the project are to improve the fertility of the soil and protect it against erosion. The principal project activity is the production and distribution of tree seedlings for planting by farmers.

Six large nurseries have been established. These are 5-10 kilometres apart and are intended to serve all villagers in the project area. They are managed by the project staff with casual labour being paid for through a food-for-work scheme. About 1.5 million seedlings were produced in 1988.

The majority of seedlings leaving the nurseries are used in village plantations. These are owned by the Peasant Associations, but are, in effect, established and managed by GLUP. The construction of microcatchments, tree planting, guarding, weeding and other operations in the plantations are all carried out on a food-for-work basis. Transport of seedlings from the nurseries to the planting site is provided by the project, which has five vehicles at its disposal.

Smaller numbers of trees are also planted by individual farmers without any food-for-work incentive and it is not uncommon to find a few dozen trees growing in a family compound. Most of these seedlings are given free to farmers, though some of the more popular species are sold at a subsidised rate. Coffee plants are the most in demand, followed by fruit trees, and then construction wood species. Newly introduced leguminous trees like *leucaena* are not popular. The seedlings are transported by project vehicles to the villages for further distribution, or are collected by the farmers at the nurseries.

The administrative structure in Ethiopia is quite authoritarian, which has both advantages and disadvantages for a project like GLUP. Balancing demand and supply of seedlings, for example, is not difficult for the project. If there is a surplus of some species, it is always possible to have the trees planted through a food-for-work scheme. Alternatively, extension workers can be sent out to tell villagers that more trees need to be planted.

Many of the plantations are situated in badly eroded and over-grazed lands. The arrangement with the local villages is that the project pays for micro-catchment construction and planting through food-for-work while the village accepts responsibility for keeping livestock out of the planted area. This has, so far, worked well, but it remains to be seen whether the villages will respect the planted areas in the longer term.

One problem the project faces is that of "nursery specialisation." This is because of the structure of the Ministry of Agriculture in which one department is responsible for coffee, another for fruit trees, and yet another for forestry trees. This is reflected in the nurseries, some of which only produce one type of species. At present, while the project vehicles are available, this does not pose any major problems but it may well be a major weakness after the project ends. It is difficult to imagine farmers walking to one nursery to pick up some eucalyptus, going to a second to get some fruit trees and going on to a third for some coffee seedlings.

Disappointing results with community nurseries

The project has always recognised that the central nurseries can only play a temporary role and that a strong community involvement is required if project activities are to become self-sustaining. The establishment of local nurseries

Paul Kerkhof/Panos Pictures

Villagisation during the early 1980s put extra pressure on the remaining tree cover; now very few trees are left.

has therefore been encouraged from the beginning and about 40 have already been set up. This is also part of an attempt to move the project away from reliance on food aid.

GLUP has supplied these nurseries with tools, plastic tubing for pots, seeds and, if necessary, pesticides. In some cases, wells have also been provided. All have been supported with technical advice.

The results have, however, been almost uniformly disappointing. The average production per village nursery has been only about 1,500 seedlings per year. One of the reasons for this is that Ministry of Agriculture staff, rather than encouraging local initiative, have provoked resistance by trying to force the villages to create nurseries. Villagers also do not see the point in setting up their own nursery when they can get most of the seedlings delivered free from the central nurseries. The fact that there are no food-for-work arrangements for community nurseries, as there are for many other development projects, also makes them unattractive to local people.

The project is examining ways in which the promotion of community nurseries might be improved. One strategy being considered is to provide them with the means to grow the more valuable seedlings such as coffee and fruit trees which they are unable to produce at present. The possibility of letting the nursery become the responsibility of an interested group in the village rather than the whole community is also being examined; this is, in other words, a possible step in the direction of privatisation.

Conflicting views on species selection

The project objectives of soil improvement and protection required the use of leguminous species like *Leucaena leucocephala*. Project extension staff were taught how such trees improve plant yields and which species are best planted for the purpose.

The project, however, had neither the time nor the resources to carry out a research programme. Instead, the usual "success package", in which *Leucaena leucocephala*, *Calliandra calothyrsus*, *Sesbania sesban* and other nitrogen-fixing trees are used for intercropping in food crops, was promoted.

The results were not at all impressive. *Leucaena*, in particular, and despite its reputation, showed an abysmal growth. It was a painful lesson in the risks of putting too much trust in the textbook. One project worker said, "I was devastated when *leucaena* didn't work. We inoculated it, we did everything to make it grow, and it was like we were growing a dwarf variety".

The same project officer said that farmers also resisted the project choice of species. "The local communities showed a tremendous interest in eucalyptus trees. This really bothered the project personnel. Some were convinced that the farmers were not expressing their needs correctly. But obviously they were. It is now accepted that until you can supply their felt needs for poles they will not be interested in multi-purpose trees."

Paul Kerkoff/Panos Pictures

Extensive terracing has been carried out as part of various food-for-work schemes.

Extension techniques

When it began in 1986, the project selected 25 men and women and trained them to become extension workers. All have had secondary education and in some cases have had higher education, up to university graduate level. They supplement the 14 Ministry of Agriculture extension workers in the project area, but are employed by CARE.

Initially they were supervised by field officers from the Ministry, but this was found to be ineffective. The Ministry's extension staff tends to be used for enforcement, which the project management wants to avoid. Subsequently, the project employed its own field officers from among its extension staff.

The extension workers received an initial six weeks' training in agroforestry and related matters, through courses given by senior project staff and visiting trainers. They also gained experience working as enumerators in some project surveys. At present, extension workers receive about 10-15 days' training per year, which is considered too low by the project management.

As extension aids, the project has used slide projectors with solar-charged batteries. These cost about US$300 each and are purchased from the international organisation, World Neighbours. They have so far turned out to be reliable and foolproof.

Slide shows are organised and presented by extension workers who have been trained to generate a response from the audience. The topics covered may vary from *leucaena* to drinking water contamination. A local community building may be used for the shows and in a typical case about 20 villagers turn up. At the end of the session some last questions, answers, and comments are exchanged.

The slides used by the project have been purchased from World Neighbours and they are probably the weakest link in this extension technique. A series showing how interactions between rhizobia bacteria and the roots of a *leucaena* lead to giant growth of the tree is pointless, and indeed counter-productive, if the species only reaches a dwarf size in Gursum. Furthermore, few villagers can be expected to grasp the concept of rhizobium-host plant interaction.

Other slide series are perhaps more appropriate, but they have the major drawback that they have not been locally produced and are often not locally applicable. Project staff are now trying to produce slide shows based on local shots which could be an important step in making the extension material more convincing and influential in changing behaviour.

The project has monitored the number of people coming to the slide shows and found that they are extremely popular among farmers. Over 15,000 attended in 1988 alone. But whether the shows bring about a higher level of awareness and changed behaviour, or are simply a form of amusement, still needs to be monitored.

The question of sustainability

Development projects face formidable obstacles in Ethiopia. The highly centralised administrative structure and the unstable security position make it extremely difficult to promote the degree of genuine community involvement required if programmes are to be self-sustaining in the long term.

In its implementation strategy, CARE has chosen to by-pass some of the bureaucratic bottlenecks by setting up its own extension service to work in parallel with that of the Ministry of Agriculture. This has undoubtedly enabled the project to have a wider and more rapid impact.

The project staff are relatively well trained and educated. They are also well paid and are given considerable means to carry out their work. Motor-cycles are available to most extension workers, field officers have cars at their disposal, and extra lorries are put to work for seedling transport in the rainy season.

On the other side of the balance, however, it is clear that the project has so far contributed little in the way of local institution-building. The training provided for the project staff contributes to the general manpower development of the country, but there is no framework within which their skills can continue to be exercised if the project funding ends. In that case, the responsibility for continuing the programme would rest with the Ministry of Agriculture, a task which, with its scarce resources, it may not be able to assume.

The dependence of the programme on food-for-work is another weakness. Food-for-work tends to have an immobilising effect on community development and means that work is carried out simply to obtain food irrespective of whether people believe in its relevance. Under such conditions, it is extremely difficult to establish tree growing and soil protection activities on a self-sustaining basis.

The project, however, must be seen within the general Ethiopian context. Most programmes in the country are based simply on the provision of immediate necessities such as food, water and medicine. Genuine development projects are rare and sustainable ones even more so. GLUP is a relatively young project that has, at least, managed to replace food relief with food-for-work.

In addition, it has involved local communities in the design of food-for-work schemes. It has also made a beginning with more sustainable activities. Tree planting and seedling production are now done to some extent without a food aid incentive. Extension has made a beginning in increasing the awareness and self-confidence of the villagers. In the longer term, this will contribute to people becoming more self-reliant.

References:

BUCK, L. et al (1986). "Gursum land use proposal". CARE.

CARE (1987). "GLUP: Third trimester report".

CARE (1988). "GLUP: Second trimester report".

PROJET BOIS DE VILLAGES,
Burkina Faso and Mali

VILLAGE WOODLOTS ARE NOT ENOUGH

Village woodlots have been widely promoted throughout Sahelian West Africa over the past decade. This attempt to involve local communities in tree growing emerged as a response to the failed forestry plantations of the 1970s.

Village woodlots represent a step forward in community participation. They permit local people to have a greater say in their own affairs than is usually the case with forestry plantations. But ironically, this improved communication with local people is revealing that village woodlots tend to be a low priority.

Accelerating loss of natural woodlands

Burkina Faso is a landlocked country with an area of 274,000 square kilometres. The climate ranges from Sahelian in the north, where there is an annual rainfall of barely 400mm, to Sudanian in the south where the rainfall is 1,000mm.

In the drier areas, millet is the major crop, together with peanuts, cowpeas, sorghum and some vegetables; the major tree species are baobab, *Khaya senegalensis*, *Acacia albida* and other *acacia* species, and various *Combretum* species. Towards the south the variety of food crops and trees increases.

The country is densely populated by Sahelian standards with a total of about 7 million inhabitants. Most people are settled farmers though there are some herders, like the Fulani. Several million Burkinabe (the people of Burkina Faso) have emigrated to the much wealthier Côte d'Ivoire in search of work and better living conditions.

Concern at the loss of natural woodlands has been growing since the 1960s. Surveys have shown that forest and woodland covered about 56% of the country in 1976 but that this had fallen to 40-45% by 1988. In the 1970s, the loss of the natural forest was widely attributed to cutting trees for fuelwood; many people saw it leading to creeping desertification and a major crisis in domestic energy supplies.

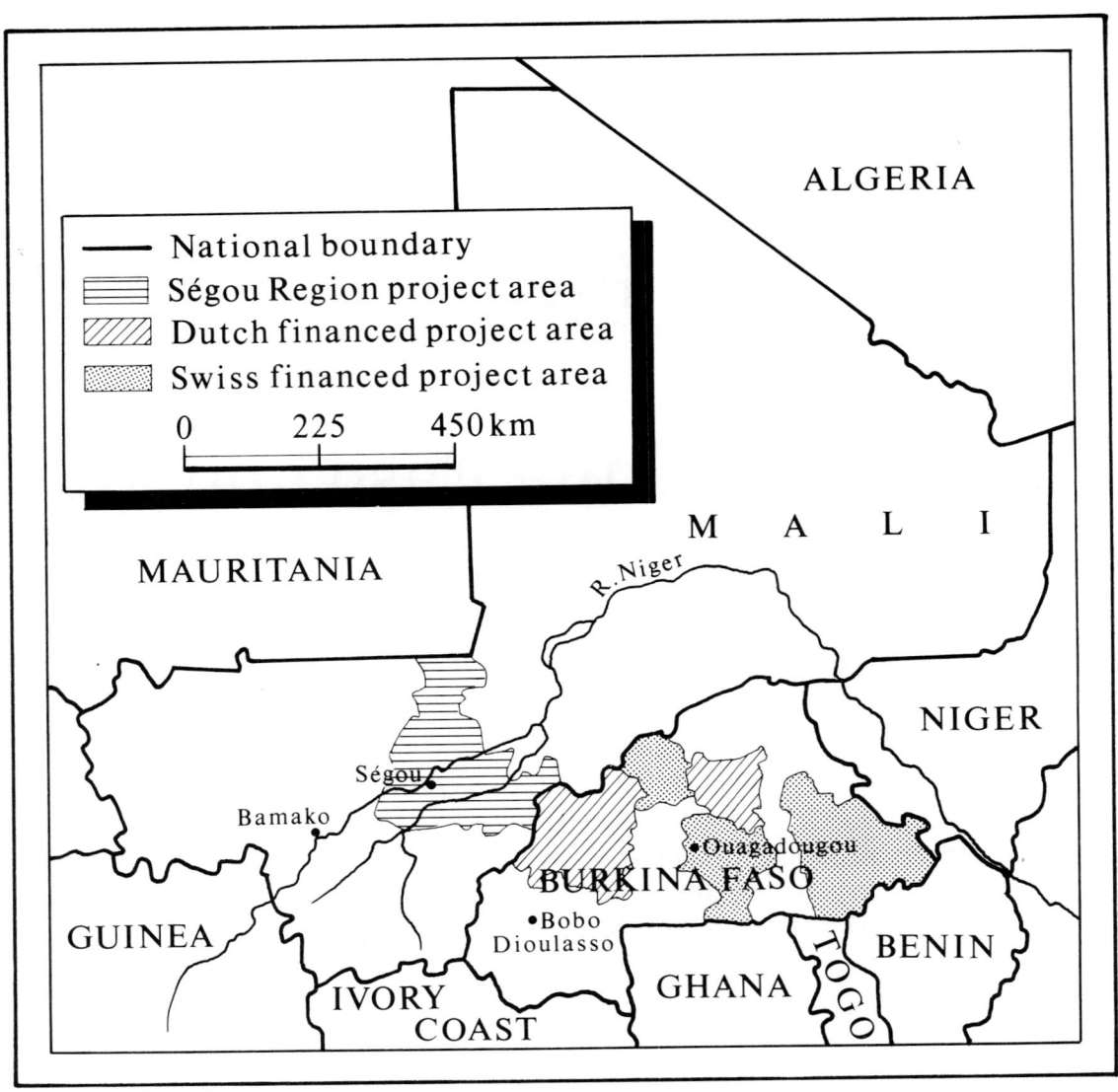

Name of project:	**Projet Bois de Villages (Village Woodlot Project)**
Addresses:	**Burkina Faso: Swiss funded: BP 2736, Ouagadougou. Tel:307389**
	Burkina Faso: Dutch funded: BP 6256, Ouagadougou.
	Mali, Direction Regionale des Eaux et Forets, BP 87, Segou.
Project area:	**16 provinces in Burkina Faso, Segou Region in Mali**
Average rainfall:	**400 - 1,000 mm per year**
Implementation:	**Forestry service with advisors**
Funding:	**Burkina Faso: combined Swiss/Dutch funding currently US$4 million per year** **Mali: total Dutch funding (1983-87) US$2.5 million**
Exchange rate:	**300 CFA = 1 US$ (early 1989)**

The village woodlot project

In the early 1970s, it was decided that a major effort to create forestry plantations was necessary. Over 16,000 hectares of plantations were established, mainly around the capital city Ouagadougou. The cost of creating and maintaining these, however, turned out to be extremely high and the Forest Department began to look for better afforestation strategies. In 1977, it began a programme to establish communal village woodlots as an alternative to plantation forestry.

The village woodlot programme is credited with being the first of its kind in West Africa. It was funded by the Swiss Government and in 1979 the Dutch Government provided support to a very similar project. Since the two have the same title (Projet Bois de Villages), and funding and evaluations have been closely coordinated, both are included in this profile.

The communal approach was decided upon because the Burkinabe peasants live mainly in villages rather than separate homesteads. The woodlots are typically a few hectares in size and the trees are planted at a spacing of 4x4 metres. The Forest Department supplied free seedlings from its own nurseries, of which there are now 47 in the project area; it also provided free fencing. All the necessary labour was supplied free by the village. The ownership of the woodlot normally rests with the men of the village. In the early years of the project, there was a heavy reliance on *Eucalyptus camaldulensis*.

The woodlots can be seen, in effect, as small versions of the large plantations established around Ouagadougou, but using village land and voluntary labour. One of their attractions for the Forest Department was that their relatively large size provided economies of scale in fencing costs. The Forest Department intention was that every village would plant a woodlot every year.

This did not happen. Most villages were only prepared to try one and leave it at that. Nor were they enthusiastic about maintenance; tree survival rates varied but were often well under 50%. The growth rate of the trees has also tended to be disappointing.

The lack of enthusiasm from the villagers forced the project to reconsider its approach. The Forest Department began to promote village nurseries as a means of creating greater local interest and involvement, as well as reducing reliance on the central nurseries. The intention was that the village

The recurrent droughts in the Sahel have caused widespread suffering; village woodlots were seen as one of the ways of fighting desertification.

nurseries would offer a wider choice of species than the few exotics provided under the project up to then. Seedlings were also made available to individual farmers for planting on their own land.

By 1985, eight years after project commencement, about 1,400 of the country's 7,000 villages had been reached in one way or another. A total of around 3,500 hectares of village woodlots had been established under the project (Swiss and Dutch finance combined) and 5-10% of the trees leaving the nurseries had gone to individual farmers.

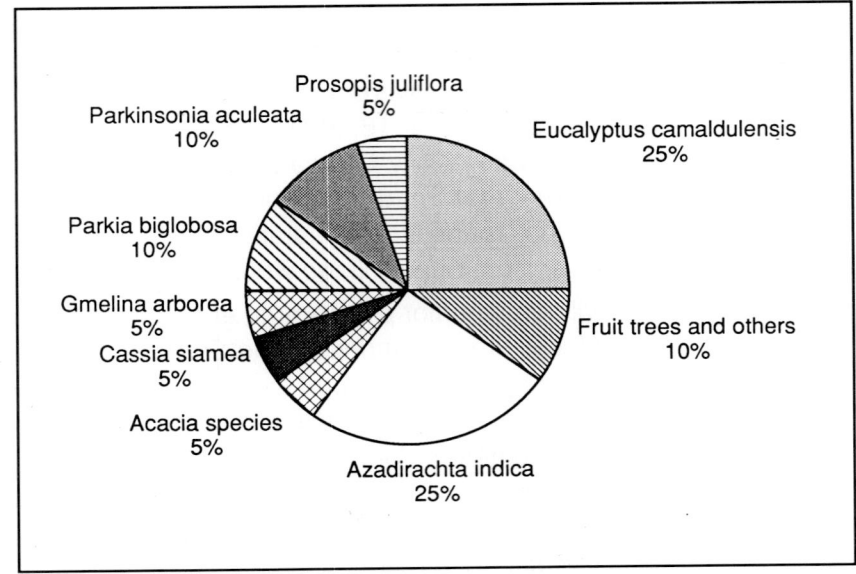

Figure 9.1 The share of different species in total nursery output (1986).

Present activities

The project is now in Phase V. Each phase has been marked by a major expansion in activity. Funding is still provided by the Swiss and Dutch governments and the total annual budget is about US$4 million per year. In 1989, there were over 20 expatriate advisers, most of whom worked for the Dutch component. Despite the heavy expatriate presence, the Burkinabe have the main responsibility for the project funds.

The principal project activity is still the promotion of village woodlots and other tree planting by villagers. Funding is also being provided for reinforcement of the Forest Department infrastructure, management of natural forests, and dissemination of improved stoves; Dutch funding is also being provided for some peri-urban plantations.

The mix of tree seedlings being provided has been widened considerably. Rather than focusing primarily on eucalyptus, a variety of other exotic and indigenous species are being produced. The division between species in 1986 is shown in Figure 9.1.

Eucalyptus and neem account for 50% of the total. This is because they grow quickly and produce good construction wood which can be sold. For this reason they are popular among the men who have responsibility for village woodlots, and among farmers growing their own trees.

Attempts to decentralise seedling production continue. The number of village or school nurseries has steadily increased but there are still numerous obstacles to be overcome. Many villages, for example, lack a reliable water supply and a considerable proportion of the nurseries are poorly managed. By 1988, the

Jeremy Hartley/Oxfam

A fenced village woodlot in Burkina Faso; only small areas have been planted in most villages and the results have often been disappointing.

village and group nurseries still accounted for only 1-15% of the total seedling production in the project.

The project is also trying to reduce the costs of the woodlots and persuade the villages to accept a greater degree of responsibility for establishing and maintaining them. The free fencing materials which have been a major cost element are to be stopped entirely. It is not yet clear what effect this will have on woodlot establishment and the long-term survival of seedlings.

Efforts are also being made to increase the amount of tree planting by individuals. In 1987, a quarter of a million seedlings were planted by farmers on their private lands and the survival rate was reported to be better than in village woodlots.

Despite the achievements to date, project staff realise that much more is needed before a significant impact can be made in the fight against desertification. It is now recognised that the loss of forest cover as a result of fuelwood cutting is relatively minor in comparison with the effects of clearing for agriculture, which accounts for an annual deforestation of 50,000 hectares; in addition, bush fires destroy a further 50,000 hectares each year. As one project staff member said: "A few woodlots are not going to stop the desert."

It has also become clear that village woodlots are an expensive way of producing poles and construction wood. Some preliminary data suggest an annual wood production of 1-2 cubic metres per hectare for *Eucalyptus camaldulensis*. Given the labour inputs and the costs of fencing, seedlings, transport and other overheads, it is highly questionable whether the woodlots are economically viable. In the circumstances, it is perhaps not surprising that the villagers have been hesitant to plant a great deal more.

Such considerations have led the project to conclude that its range of activities needs to be broadened considerably beyond tree planting. In future, much more attention will be paid to the agricultural system. Natural regeneration and tree planting in crop fields will be considered as well as the potential for mechanical soil conservation measures.

The GRAAP method in action; an extension worker in Burkina Faso uses pictures on a flannelboard in discussions with local villagers.

The use of flannelographs

When the project started in 1977, the Forest Department's main work was in plantation forestry and forest protection, and it had virtually no experience of working with villagers. An extension system therefore had to be created from scratch; in doing so, the Department collaborated with GRAAP, a social research institute in Bobo Dialassou in southern Burkina Faso.

GRAAP believes that rural people have lost faith in their ability to influence their own future and argues that it is necessary to promote discussion among villagers themselves so that they can understand the changes taking place in their environment. Then, GRAAP says, they should be given the chance to decide themselves what action to take. The role of extension agents is to promote and facilitate these discussions.

According to GRAAP, communication in the rural African setting typically takes place under a big tree, where the villagers meet and discuss issues and events. Verbal discussions in this setting, rather than radio, television or the written word, should be used by development projects when they wish to interact with villagers. This process can be facilitated by a project extension agent. The use of cut-out pictures on a flannel board, so-called "flannelographs", are used to make the work of the extension agent easier and to stimulate discussion (GRAAP, 1984).

The agent makes a presentation using pictures which show what the village environment was like in the past. The villagers at the meeting then split into working groups according to sex and age. These discuss the presentation and reach their own conclusions on how things were in the past and why. There is then a plenary session in which a general consensus is reached.

The process is repeated for present conditions and the way events are likely to turn out in the future. The final step is to agree on what should be done about the problems identified by the villagers. Leaflets or booklets are distributed by the extension agent to remind the participants of the major issues discussed.

During the discussions, dubious or dissenting arguments are dealt with by the extension agent. For instance, the old men may argue that: "These days the rains fail because the youth no longer show any respect." The agent may respond with the question: "But why is it, then, that the rains are abundant in Ivory Coast when everybody knows that the youth in that place show no respect at all?" The old men will agree that their argument is dubious and the discussion continues until a consensus emerges.

The flannelographs used by the extension agent are adapted to the local environment and culture. GRAAP argues that there is little point in showing the peasants microscopic pictures of bacteria, Rhizobium and suchlike modern representations. The pictures should instead resemble local art and be close to local cultural expression. The representation of a third dimension in figures and drawings, for example, often makes no sense to villagers. This has led to the design of the typical two-dimensional GRAAP drawings.

The GRAAP method has now been in use for about a decade in West Africa. There have been no measurements of its effectiveness to date in the village woodlot project but it is not without its critics. One view is that it is condescending and paternalistic. Such critics argue that African peasants are not so ignorant that they do not know their own problems or what is required to solve them. Their real need is for technical and financial help rather than an agent helping them to formulate what they already know.

Another criticism is that, once it has been established what the villagers really want, they are not given it. According to GRAAP: "To work with the peasant one has to depart from where he lives, what he knows, what he can do, and what he wants." The villagers are encouraged to talk about education, livestock disease, lack of water and lack of land. But when it comes to action, the extension agent, who has no mandate for anything else, can only tell them to plant a woodlot.

Finally, the GRAAP method assumes that the extension agents using it are well trained and open minded. If they are not, as is the case with some foresters, rather than stimulating a dialogue among the peasants, the session can turn into a lecture on how to establish a woodlot.

The village forester

The GRAAP method played a large part in the early extension efforts of the project. It helped raise local awareness of environmental issues and was also important in opening communication between foresters and villagers. But once it had been used a few times, villagers began to "know it by heart" and other approaches had to be developed.

The most important of these was the idea of the village or peasant forester (paysan forestier). This is someone who is chosen by the village to be trained in tree growing, nursery management and other relevant techniques in order to be available to villagers whenever they need advice and help.

Forest Department agents are responsible for stimulating villages to hold a general meeting in which the tasks of the village forester are discussed and defined and someone is chosen to fill the role. This can only be done in villages which have already discussed their environmental problems and are convinced of the relevance of tree growing to ·solving them.

The intention is that the village forester becomes part of a process by which villages increasingly accept responsibility for the management of their own natural environment. Such a development is inevitably slow and there are few signs that it has yet happened to any significant extent. Nevertheless, the village forester represents a major step away from the oppressive attitudes, inherited from the colonial regime, which have tended to characterise relations between local people and the Forest Service.

Paul Kerkhoff/Panos Pictures

One of the more successful village woodlots in Mali.

Replication in Mali

The village woodlot project of Burkina Faso has been used as the model for many other projects in West Africa. The village woodlot project of Segou in central Mali is one example.

On the basis of two fuelwood surveys, one in Bamako and one in the project area, it was concluded that fuelwood shortages were a serious local problem. A village woodlot project, based on the Burkina Faso model, was designed and started in 1983. The Dutch government provided funding of US$2.5 million over the period 1983-87 and the Forest Department was the implementing agency. The GRAAP method was used for communication with the local people.

Some five years later, the project finds that its initial analysis of local needs and priorities was inadequate. It is now recognised that the local people do not

94

see fuelwood shortages as a high priority problem. Neither are they particularly interested in communal woodlots, and those which have been planted often have poor survival rates and low production (Deneve et al, 1988).

The project is now trying to diversify its activities, but this is a slow process. The woodlot approach represents traditional forestry thinking applied in a village context. If the project is to go beyond this, a whole set of attitudes will have to be changed within the Forest Department.

Learning the necessary lessons

Several lessons can be drawn from the village woodlot experience in West Africa. One of the most important is a recognition of the distortion in development priorities which occurred as a result of what has been called the "fuelwood syndrome". This was the belief that fuelwood cutting was the main cause of deforestation which in turn was leading to widespread desertification and an energy crisis. The result was that there was no proper diagnosis of local needs; it was simply assumed that people were interested in planting village woodlots to produce fuelwood.

In practice, this meant that a great deal of time and effort was expended on creating woodlots which have turned out to be costly and unproductive as well as irrelevant to the problem of environmental degradation. They also exclude important social groups like women, herders and migrants.

There have, however, been some positive results. The woodlot programmes have had a major impact on Forestry Department thinking, at least in Burkina Faso. Thirteen years ago foresters saw themselves as members of a forest police force. Now they act much more as extension agents, trying to assist villagers rather than simply fining them. It is also probable that the existence of village woodlots has changed the attitude of local people towards growing trees.

There is, however, no disagreement on the need for a broader approach in the future. If there is to be effective action against desertification, it cannot be based simply on forestry plantations or the promotion of village woodlots. Local populations will have to be provided with the techniques and incentives to become involved in the management of their own natural environment.

Above all, programmes will have to address the problems of the agricultural sector. Whether Forest Departments are necessarily the best institutions to develop such innovations is another question.

References:

DENEVE, R. et.al (1988). "Rapport d'évaluation du projet 'Bois de Villages, Ségou' au Mali, Financement Pays Bas".

GRAAP (1984). "Pour une pedagogie de l'autopromotion". 4ème edition actualiseé. GRAAP, BP 785, Bobo-Dialassou, Burkina Faso.

KABORE, V. et al (1985). "Rapport d'évaluation des projets 'Bois de Villages' Néerlandais et Suisse".

THE VILLAGE AFFORESTATION PROGRAMME, Tanzania

GROWING THE NATION'S FUEL

Tanzania's Village Afforestation Programme has a history of over two decades. It began in the late 1960s as part of a nation-wide reafforestation effort. In the 1970s, it came to be motivated by a belief that the country faced a massive woodfuel deficit.

The programme is now oriented much more towards pole and timber production and the communal approach is giving way to the promotion of tree growing by individual farmers.

The early beginnings of the programme

The Arusha Declaration in 1967 marked a major change in the political direction followed by the country since its independence in 1961. It placed an emphasis on socialism, decentralisation, and communal self-help. The Village Afforestation Programme was designed in the spirit of the Declaration and was intended to play a major part in the country's afforestation efforts. It is planned and implemented by the Forest Division of the Ministry of Natural Resources and Tourism. Support has been provided by agencies such as SIDA, FAO, ILO and DGIS, with the main funding coming from SIDA.

In the 1970s, attention began to focus on the issue of rural energy supplies. The Forest Division estimated that the annual consumption of fuelwood in the rural areas was 2 cubic metres per head. Using this as a basis, calculations showed that the total consumption of fuelwood was greatly in excess of the estimated wood production of the country. This led to an increasing emphasis on village woodlots as a means of meeting future fuelwood shortages.

The main promotion method was to supply villages with seedlings which were produced and delivered by the Forest Department. Political pressure was often placed on the village leadership to ensure that land was set aside for the plantation. Sometimes, the planting was done by the Forest Division, at other times by the villagers.

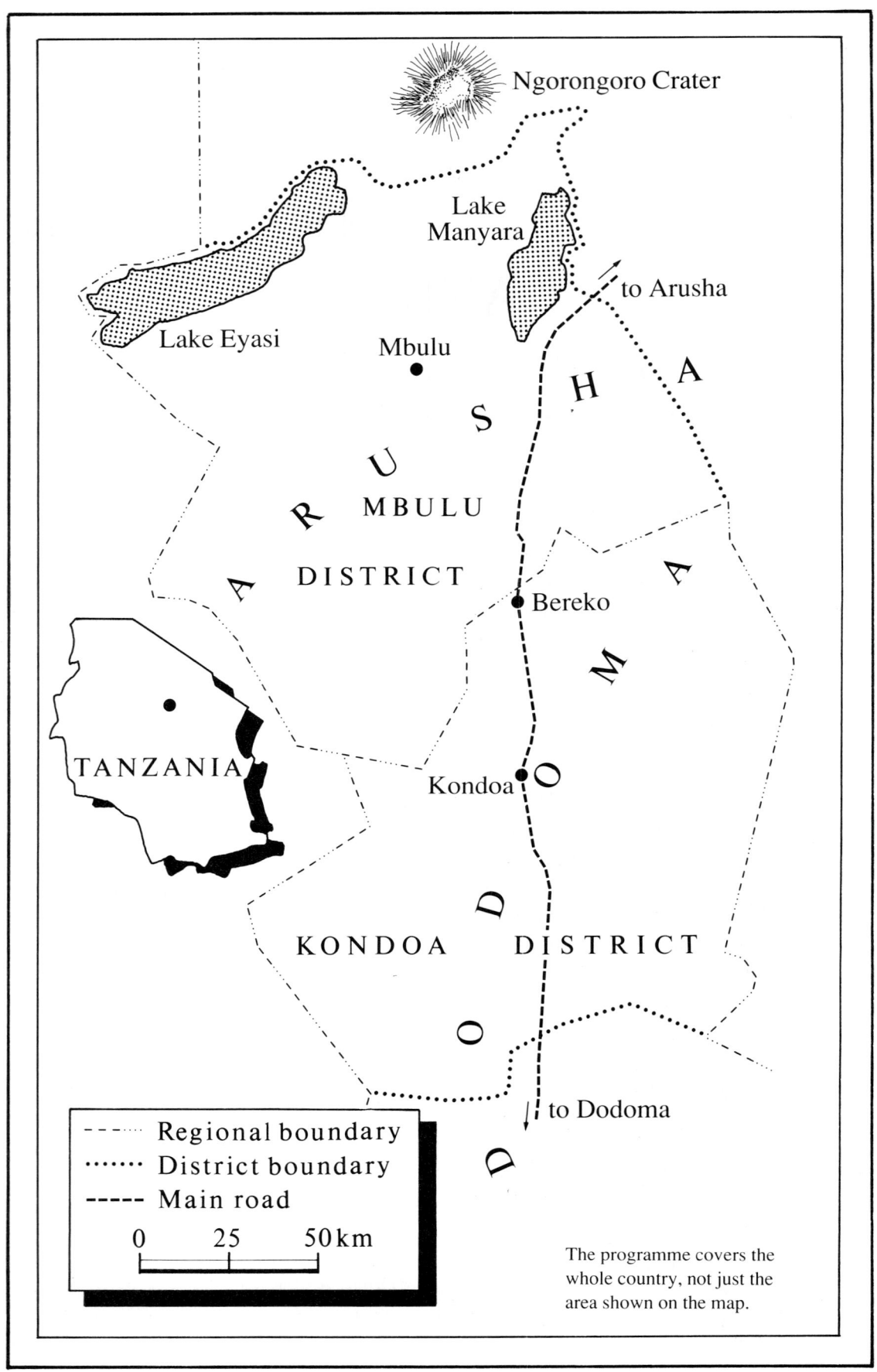

Ngorongoro Crater

Lake
Manyara

to Arusha

Lake Eyasi

Mbulu

A
R
U
S
H
A

MBULU

DISTRICT

Bereko

D
O
D
O
M
A

Kondoa

TANZANIA

KONDOA   DISTRICT

D
O
D
O
M
A

to Dodoma

- - - - - Regional boundary
· · · · · · District boundary
- - - - Main road

0   25   50 km

The programme covers the
whole country, not just the
area shown on the map.

Name of project:	Village Afforestation Programme
Address:	The Forestry Department, P.O.Box 426, Dar es Salaam, Tanzania.
Project area:	nationwide programme
Average rainfall:	500 - 1,500 mm per year
Implementation:	Forestry Division with expatriate advisors
Funding:	SIDA, FAO, ILO, DGIS
	Average funding level (1985-89) US$ 800,000 per year
Exchange rate:	120 TSh = 1 US$ (late 1988)

The total annual planting needed to meet projected fuelwood demands was calculated to be about 400,000 hectares per year. The rate of planting, however, never remotely approached this and towards the end of the 1970s it was still well below 10,000 hectares per year; and even this is an optimistic figure as it was based on the number of seedlings distributed and assumed a high survival rate.

Several reasons for the limited impact of the programme at this stage have been suggested. One was that village forestry was the responsibility of the Forest Management Section of the Forest Division. This was completely lacking in experience of extension work and saw its task as simply telling villagers how to establish forest plantations. An extension film of the time, for example, explains that village woodlots should be planted with conventional forestry species such as *Pinus* and *Cupressus* at 2.5 metre spacings.

By the early 1980s, the Forest Division was convinced that a change in direction was required (Mnzava, 1983). A separate Village Forestry Unit was set up to develop new promotional methods. It produced a number of publications for foresters and extension workers in which a more participatory approach was emphasised. Although there were villages, however, where woodlots were established and managed satisfactorily, the overall rate of tree planting still did not show any significant improvement.

Lars Johansson/Panos Pictures

*A village committee meet in the shade of a flourishing nine-year-old **Cassia siamea** woodlot; some woodlots have been succcessfully established, but many more have failed.*

Review of basic assumptions

Several studies were carried out to try to establish the reasons for the relative lack of impact of the programme. These revealed a number of fundamental flaws in the analysis on which the programme had been based.

One of its basic premises had been that Tanzanian peasants were facing a fuelwood crisis which had to be solved by tree growing.

In fact, the farmers placed a very low priority on producing fuelwood and were generally much more interested in growing trees for construction wood, poles, fruit or other non-fuel purposes. Moreover, it was found that the figure of 2 cubic metres per head per year, seriously exaggerated fuelwood consumption in many areas. An FAO study, for example, found that consumption varied between 0.5 and 1.5 cubic metres per head per year, in a series of 15 villages (FAO, 1984).

Distribution of seedlings from the large central nurseries in which they were produced was found to be a serious problem. In fact, given the lack of resources at the disposal of the Forest Division, it was clearly impossible to distribute such nursery production to the thousands of villages in the country.

Paul Kerkhof/Panos Pictures

This village woodlot — in the foreground — failed completely because it was planted too close to a cattle track.

The degree of collaboration with villages was also found to be insufficient in many cases. Foresters were under considerable pressure to produce tangible results and, at times, simply used the Division's labour, tractors, fencing and other materials to produce large plantations, some up to 50-80 hectares, which were then termed "demonstration woodlots".

There was little or no village involvement in these and their demonstration effect was probably negligible. At times, there was even conflict, with villagers wanting to harvest produce from the plantations and foresters claiming they belonged to the Division.

Perhaps the most important realisation was that the communal approach was profoundly unpopular among the majority of villagers. People turned out to be distrustful of the village leadership in many areas. There were instances where village chairmen were said to have sold the produce of a woodlot and kept the money, or used the wood to improve their own dwellings. In many cases, there was uncertainty about what to do with the woodlot once it had been established and how to allocate responsibility within the village for managing it (Skutsch, 1985).

The failure of the communal woodlot policy is now extremely obvious. An untended woodlot in front of the village headquarters is a common sight; livestock roam freely through it because there is no one to keep them away. Reports of tree seedlings being stolen the night after they have been planted are not surprising given that people do not identify with the woodlots or feel they belong to them in any way. At the same time, it shows that there is a demand for tree seedlings.

Experience in Kondoa District in Dodoma Region

The Forest Division in Kondoa District became actively involved in the Village Afforestation Programme in 1973. It relied on a highly centralised approach

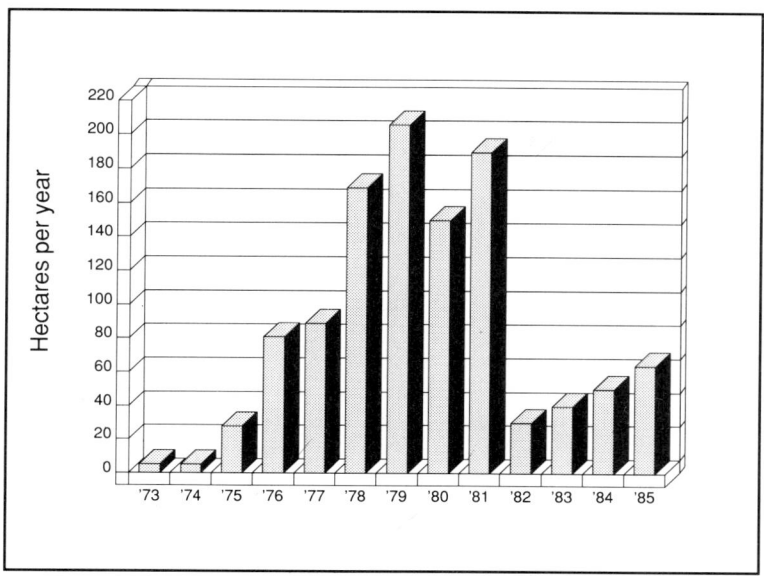

Figure 10.1. Area of Woodlots planted in Konda District (1973-85).

and produced all the seedlings in its own nurseries. It also carried out the clearing, digging, planting, weeding and application of pesticides. The village supplied the land and gave a certain amount of assistance in the establishment of the plantation.

The area of plantations created in this way each year during the period 1973-85 is shown in Figure 10.1. As can be seen, there was a rapid increase during the first seven years up to a maximum of 206 hectares in 1979. In the early 1980s, however, the area planted declined rapidly and by the 1988/89 season it was virtually nil. This rise and fall is reflected in the change in thinking behind the programme which took place during this period.

In the early years, when a communal woodlot was being planned, the foresters approached the village leadership who were usually quite positive about the idea; the creation of village woodlots was seen as an important national issue. But individual villagers were rarely consulted either by the foresters, who had to cover the whole District, or by the village leaders.

After the woodlot had been established there was usually a slackening of official interest. The emphasis in the programme, at this stage, was primarily on meeting centrally-set targets and there was little provision for follow-up or long-term maintenance.

This meant that villagers were not clear about the purpose of the woodlot or who was supposed to benefit from it. There was therefore little interest in protecting or managing the majority of woodlots, and damage from grazing, burning and cutting was widespread. Once what was happening became clear, foresters lost their enthusiasm for promoting the woodlots and this is reflected in the decline in the area planted each year.

The Bereko Division of Kondoa District where an ILO-funded project has been running since 1985 was, however, somewhat different. One of the major aims of the programme in this area is the creation of employment. This is done by using local labour for the establishment of the woodlots and has generated considerable local enthusiasm for the programme. Each village is also provided with a paid Forest Attendant, which has greatly improved the care and maintenance of the plantations.

None of this, however, means that there has been any fundamental change in the prospects for the long term survival of the woodlots. As one forester said: "The programme is successful in Bereko because people are paid to do the work. As long as the funding lasts, the woodlots will continue to be established and protected, but when it stops everything will finish."

The Forest Division now prefers to promote tree planting by farmers on their own land. It is distributing about 200,000 seedlings per year, most of which go to individual farmers and schools. In dry areas the seedlings are distributed free but in areas of high agricultural potential they are often sold.

Seedling production is still, however, highly centralised with only five nurseries to serve the whole of the Kondoa District. These depend on the Division's motorised transport for seedling distribution which means that only a minority of villages can be reached every year. There have been attempts to promote small community nurseries, but without much success so far. Bad soils and lack of water are the major technical obstacles.

One year, the difficulties in getting transport at the beginning of the rainy season were so great that seedling distribution was impossible. The Division discussed this with the District Development

Lars Johansson/Panos Pictures

The programme is now putting more emphasis on tree planting by individuals; a heavily protected **Cassia siamea** *seedlings planted in a family compound.*

Director, pointing out that it might as well stop raising seedlings if they could not be distributed. Since then, the necessary transport has always been provided, if necessary with hired vehicles.

One of the biggest obstacles facing the Forest Division in Kondoa District is the lack of trained staff. It has only about 20 extension workers to cover the whole District. Most have only attended primary school and some have had no formal education. They are not trained to approach farmers or address meetings and some can hardly be called extension workers. This makes it difficult for the small number of foresters in the District to do much beyond seedling production and distribution.

Experience in Mbulu District in Arusha Region

Mbulu District borders the famous Ngorongoro Crater in Northern Tanzania. It is a mountainous area with an altitude of 1,500 to 2,000m or more. The agricultural potential is considerable and there is a high population density. Livestock rearing is traditionally a major economic activity but crop production is increasingly important.

The Forest Division in the district is supported by SIDA and also, more recently, by DGIS of the Netherlands. It is relatively well endowed with resources and has six foresters and 26 Forestry Assistants. The Assistants

Primary school pupils tending a school nursery; part of the Forest Division's efforts to decentralise seedling production.

have reached Standard 7, that is seven years of primary education, and some have also received three months forestry training.

There is financial support for seminars which are helpful in providing further training and increasing motivation. The seminars also act as a means of monitoring and evaluating extension work on individual farms and communal woodlots. The seedling survival counts by the Forestry Assistants, for example, are discussed at the seminars and are an important input for the next year's plans. The financial support provided by the donor agencies has also improved the availability of transport, one of the major bottlenecks in forestry extension in Tanzania.

There are nine central nurseries and the total number of seedlings distributed during the 1987/88 season was 1.5 million. Seedling production is also being devolved to a village level.

At present, there are 19 village nurseries which produced a total of 370,000 seedlings in the 1987/88 season. A further seven school nurseries have also been established. The Forest Division provides these with materials and technical advice but responsibility for their operation rests with the village or school headmaster.

The main species preferred by the farmers are eucalyptus, cypress, and grevillea. Fruit trees are also popular but are the responsibility of the Ministry of Agriculture. The Forest Division, however, intends to become fully involved in the production of fruit tree seedlings.

Experience in this District shows that village woodlots can be successfully established if there is a strong sense of community identity and the whole village council rather than just the chairman are involved. Communities also feel that if they must obey national tree growing directives it is best to have them all in one large fenced plot: the village woodlot.

But tree planting by individuals and schools is much more popular. This

Table 10.1. Comparison between tree planting in village woodlots and tree planting by individuals and schools (1986/87).

	Area planted (hectares)	Survival
Village woodlot	167	51%
Individuals	433	89%
Schools	50	n.a.

Source: Mbulu District office.

can be seen clearly in Table 10.1 which shows the relative areas established and the survival rates of seedlings. The main reason for the better survival of the privately planted trees is that they are well protected from damage by livestock. They are also weeded and ash is applied to prevent termite attack. The village woodlots, in comparison, receive little attention once the trees are planted.

From village woodlots to people's choice

It is now clear that the policy of promoting village woodlots as the basis of a national afforestation strategy has not been successful. Tree planting by individual farmers is now becoming the core of Tanzania's forestry extension.

Under the new policy, rural people are encouraged to plant trees wherever they wish. If they want to create a village woodlot, they will receive help in doing so; they will be equally helped if they want to plant trees on their own land. The production of seedlings is also being decentralised from large Forest Division nurseries to village and school nurseries. Tools, plastic pots and technical advice are provided by the Division wherever there is local interest in setting up a nursery.

The Tanzanian experience shows that, given a real interest from villagers, communal woodlots can be successfully established and there should be provision for them in any national afforestation strategy. But promoting tree planting by individual farmers for a variety of purposes, and rarely with fuelwood as the main motive, is generally likely to be much more productive.

Since it is impossible to provide such programmes with seedlings through central nurseries, decentralised production needs to be promoted. This will require continued of external support in basic materials, transport and training for a considerable time into the future.

References:

FAO (1984). "Tanzania fuelwood supply and consumption in arid areas". Forestry for Local Community Development Programme, FAO, Rome.

MNZAVA, E.M. (1983). "Tree planting in Tanzania: a voice from villagers". Forest Division, Dar es Salaam.

SKUTSCH, M. (1985). "Forestry by the people for the people – some major problems in Tanzania's village afforestation programme". International Tree Crops Journal, No.3.

REFORESTATION AROUND WELLS IN NORTHERN SENEGAL

PLANTATIONS FAIL AGAINST DESERTIFICATION

Deep wells solved the problem of dry season drinking water for pastoralists in northern Senegal. But the resulting increase in livestock concentration, combined with drought, led to over-grazing and the threat of desertification.

A project to combat this by means of large-scale tree plantations was launched in 1975. By the mid-1980s, it had become apparent that the local herders and farmers had little interest in growing trees. The project now accepts that problems of livestock management need to be tackled directly if anything is to be achieved in the area.

The Region du Fleuve

The Region du Fleuve lies between the Senegal and Ferlo rivers in northern Senegal. It is in the Sahelian climatic zone and has an annual rainfall which ranges from 200 mm in the north to about 400 mm in the south. There is a long dry season of 8-9 months and the soils are very sandy though they do not form dunes.

The region has a dry savanna ecology interspersed with riverine woodlands along the watercourses. *Balanites aegyptiaca*, *Adansonia digitata* (baobab), *Boscia coriacea*, and *Acacia seyal* are among the tree species found in the savanna and the ground cover is made up of a variety of grasses and leguminous herbs.

The temperature is somewhat reduced by a breeze from the Atlantic but reaches a mean monthly maximum of 40°C before the onset of the rains in June. When the dust storms from the Sahara reach the area, thick layers of dust settle even indoors in a matter of hours. Life there, on the threshold of the biggest desert in the world, becomes very uncomfortable.

The major part of the project area is inhabited by the semi-nomadic Fulani, or Peul, who are thought to be descended from the Nilotic herders of East Africa.

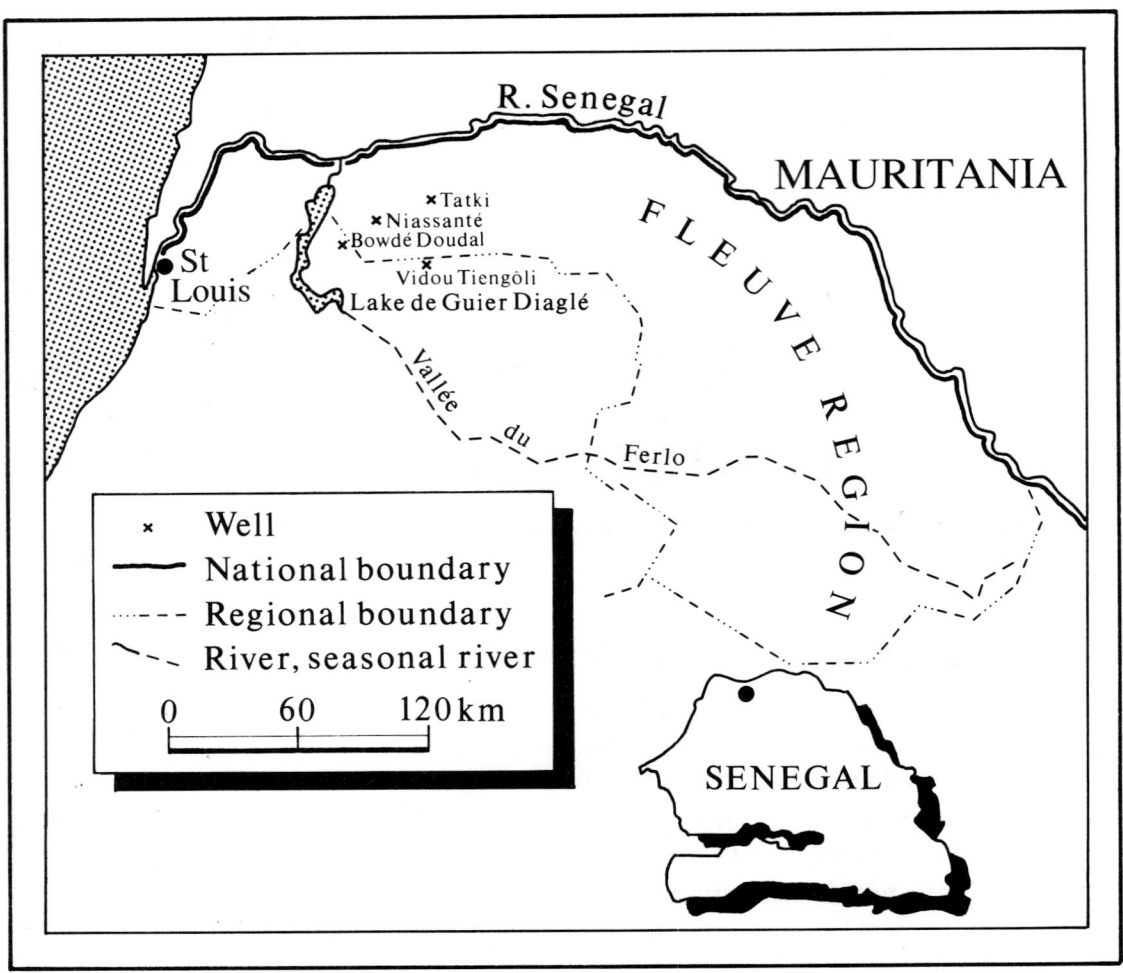

Name of project:	**Until recently: Reforestation around wells in northern Senegal**
New name:	**Agro-sylvo-pastoral landuse models in the fight against desertification**
Address:	**Mission Forestier Allemand, BP 366, St.Louis, Senegal. Tel: 611120**
Average rainfall:	**300-400mm per year**
Project area:	**Region du Fleuve in northern Senegal.**
Implementation:	**GTZ and the Forestry Department**
Funding:	**GTZ**
	1975/79 US$2.5 million
	1979/81 US$2.3 million
	1981/83 US$3.5 million
	1983/87 US$6.0 million
	1987/91 US$7.5 million
Exchange rate:	**300FCFA = 1 US$ (early 1989)**

Crop cultivation is limited and covers perhaps 5% of the area. It is mainly practised in the south by the settled Wolof people. Millet and beans are the principal subsistence crops with peanuts being grown for cash. All these crops can be cultivated with an annual rainfall of 300-400mm provided it is well distributed during the rainy season.

Problems brought by new wells

Until the 1950s, there was no reliable all-year water source between the Senegal and Ferlo rivers. During the dry season the herders therefore retreated towards these rivers and sometimes even migrated as far as the Niger River in Mali.

To provide dry season water supplies, the colonial government constructed a series of wells which tapped into the "fossil water" found at very great depths; some of the wells were as deep as 750 metres. This fossil water was laid down 50-200 million years ago and the interchange with the groundwater is extremely slow, if it occurs at all. The total fossil water reserves are estimated to be about 5,000 billion cubic metres.

The wells are about 25 kilometres from each other and have had a major impact on the Fulani herding strategy. Once reliable water sources were available, moving to the rivers during the dry season became unnecessary. Instead, the herders settled down near the wells moving away during the rains and returning during the dry season.

With the permanent availability of drinking water, the livestock numbers began to increase, as did the human population. The population density grew to about 5 persons per square kilometre in the north and 15 persons per square kilometres in the south. The result was that, though the supplies of drinking water remained ample, dry season fodder supplies began to come under pressure.

The problem was particularly acute in the vicinity of the wells which were used daily by thousands of animals. Even in years of good rainfall, when the grass cover is excellent and young trees of the indigenous species flourish, there was heavy damage to the vegetation within 1-2 kilometres of each well.

The severe Sahelian drought of 1973-75 had a major impact. The herders were, in effect, trapped in the area as their traditional migration patterns had been broken. The grass near the wells was quickly exhausted and green trees were stripped of their branches in the frantic

Paul Kerkhof/Panos Pictures

Vindou Teingooli, one of the fossil water wells in the Ferlo region; thousands of animals are brought here during the day to drink.

search for fodder. Livestock died in great numbers and, because of the lack of rain, the regeneration of both grass and trees could no longer take place.

Reforestation around the wells

There was widespread national and inter-national concern at the destruction of the tree cover in the area. In 1975, the Government of Senegal and GTZ began a forestry project entitled "Reforestation around the wells in northern Senegal" aimed at restoring the tree cover.

It was established under the auspices of the Forestry Department, but is operated on a semi-independent basis. Its headquarters are in St Louis, about 200km from the project area and therefore. The total area covered was about 50 by 100 kilometres. The work programme was divided into phases and is currently in Phase V.

The initial objective was to fight desertification around the wells by means of reforestation. Traditional forestry techniques were used. The areas earmarked for planting were cleared of most of their remaining bush, chain-link fences were erected and the soil was mechanically prepared. The species planted was *Acacia senegal,* generally at 5x5 or 6x6 metre spacings. Guards were hired to prevent damage to the new plantations by animals or people. In addition to protecting the land, the plantations were intended to provide a source of local income in the form of gum from the *Acacia senegal.* They were also supposed to provide forage for goats.

The success achieved was extremely limited and the approach was abandoned in 1983. Of the 3,600 hectares planted, only 800 hectares were effectively in existence in 1985; and even in these the survival rate was in the range 46-78%. Moreover, the surviving *Acacia senegal* trees had not produced gum in any significant quantities. According to some project staff, natural regeneration of either *Acacia senegal* or any other tree species is virtually non-existent in the plantations.

It also became clear that the economic prospects of the plantations had been over-estimated. Calculations made in 1986 showed that the establishment costs were about US$500 per hectare whereas the anticipated benefits were only US$250 per hectare over a 30-year period. Thus even if the *Acacia senegal* had produced the anticipated quantities of gum, the plantations would not have been commercially viable. No feedback has been obtained on how the Fulani view the plantations.

Paul Kerkhof/Panos Pictures

An old, battered fence marks the boundary of a failed plantation. The plantation, on the left, was cleared of all natural vegetation before planting; now hardly any trees remain.

Attempts to involve farmers in growing trees

By the end of the first phase of the programme in 1979, some of the inherent problems were beginning to become apparent. The project staff therefore began to look for more effective strategies. Anticipating greater local interest and participation from settled cultivators, tree planting was started in the Wolof farming area in the south of the region.

Table 11.1 Costs of direct seeding

Item	Cost (FCFA/ha)
Fencing	20000
Soil tillage	10 000
5 years weeding	50 000
8 years guarding	65 000
TOTAL	145 000 ($480 US)

The villagers were asked to allocate a piece of land of 20-200 hectares for tree growing. The project then provided a chain-link fence and carried out mechanised soil preparation with a disk plough. *Acacia senegal* seedlings were brought to the site and the farmers were asked to plant them at spacings varying from 6x6m to 10x10m and cultivate their crops in between. By 1985 a total of 3,585 hectares had been planted in this fashion; under the project they were referred to as village or farmer plantations (in German, "Bauernaufforstungen").

From the project point of view, this provided a cheaper way of establishing *Acacia senegal* plantations and offered a hope of greater community involvement. Project staff assumed that the farmers would be interested in the trees as a source of fuel and that the gum would provide a cash income. It was also thought that *Acacia senegal* was a good tree for intercropping.

The step from conventional plantation forestry to a participatory approach, however, turned out to be much more difficult than anticipated. Although the project succeeded in getting the farmers to do the planting, the level of genuine community involvement was minimal and the villagers had little interest in the trees. In 1985, the tree survival rate in the best village plantations was estimated to be 40%.

There were valid reasons for the lack of local interest. Fuelwood does not appear to be a major concern locally; the gum has not been produced in significant quantities; and *Acacia senegal* is not seen as a particularly useful intercropping tree. The farmers have now stopped growing crops in the plantations which have reverted to bushland.

During the same period, the project also tried direct seeding as an alternative to planting. The approach was entirely non-participatory. Parcels of land were fenced off, disk ploughing was carried out with heavy machinery and tree seeds were sown. The parcels were guarded by watchmen. A total of 1,376 hectares were treated in this manner by 1985. As can be seen from Table 11.1, the technique was extremely costly. In the event, direct seeding was also a technical failure and was abandoned in 1985.

New directions

The first sociological study in the project history was carried out in 1985. This pointed out that contacts between project staff and the local people and village authorities were almost non-existent. Although project staff were sometimes seen visiting the plantations they did not use such opportunities to talk to people. As a result, the local population had little idea what was going on, why the plantations were being established, and who was supposed to benefit.

One of the recommendations of the study was that contacts between project staff and local people should be greatly improved. Another was that tree planting should only be undertaken when there was a significant financial contribution from the local people.

From 1985 onwards, a new project strategy was therefore followed. Local farmers are asked to form themselves into groups and provide and clear a parcel of at least 20 hectares of crop or fallow land for tree planting. The project supplies a chain-link fence and carries out mechanical soil preparation, but only after 11% of the cost has been paid for by the group of farmers. Another 11% is supposed to be paid after the harvest, since that is the time farmers tend to have more money. The total cost of the fence and land tillage is about FCFA 350,000 per hectare (US$1,100).

At the onset of the rains, the project brings the tree seedlings for the farmers to plant. Initially the project insisted on a 10x10 metre spacing and provided only *Acacia senegal* seedlings. Later a spacing of 10x20 metres was allowed and seedlings of other trees such as *Acacia albida* were provided. The farmers continue farming on the protected plots.

From the perspective of the project, the fact that the local farmers are making a substantial cash contribution is a big step towards making the project more sustainable. The total contribution was held at 22% for the period 1985-88 and was increased to 33% for 1989. It is hoped it can be increased to 80% by 1991 but there is some doubt that people will be prepared to pay so much.

A number of possible ways of reducing the costs of the fencing and soil

Table 11.2 Summary of project tree planting achievements to date		
Period	**Area (ha)**	**Type of planting**
1975-83	800	plantation forestry
1981-85	3,585	village forestry/minimal participation
1985	60	with 22% local contribution
1986	124	with 22% local contribution
1987	150	with 22% local contribution
1988	205	with 22% local contribution
1989 (projected)	120	with 33% local contribution

preparation have been suggested. Hedges of *Prosopis juliflora*, for example, have been planted on the inside of the chain-link fences in the hope that they will take over the function of the chain-link. But so far, no effective *Prosopis juliflora* hedges have been established.

The planting achievements of the programme to date are shown in Table 11.2 As can be seen, the participatory approach is a great deal slower than the creation of traditional forestry plantations. Whether this is reflected in improved and more self-sustaining long term results is, as yet, an open question.

It is also uncertain whether this project package is economically attractive (Baum, 1988). Some calculations have shown a high economic rate of return but since the greater part of the benefits are based on the assumed cash value of the anticipated fuelwood yield, such results need to be viewed with a considerable degree of caution.

Are trees the main attraction?

There has been no systematic evaluation of the farmers' feelings about the present approach. Nevertheless, the indications are that they greatly appreciate the fence which is very effective in protecting their crops. It may be that this is the reason they are prepared to make the 11%, and now 16%, initial contri- bution. The sub-sequent payment has not, however, been paid in many cases.

Whether they appreciate the *Acacia senegal* planted in their croplands is doubtful. Many farmers have already replaced these with *Acacia albida* and *Balanites aegyptiaca*. The project calculation that, because of its gum production, *Acacia senegal* is profitable over a 30-year period is not apparently persuasive at a local level.

Indeed, it is not clear whether tree planting is locally accepted as a profitable and relevant farming activity. There is a UN organisation in Ferlo that gives food to villages who show they are "good farmers". Tree planting is often seen as "good" and "progressive". Do farmers plant trees to show that they are "good" and consequently obtain food aid?

When there are incentives offered by a tree planting project — such as an expensive fence or free food — the trees themselves may be of minor importance. The current project staff feel there is a strong need to broaden the scope of the project and promote only those activities that are clearly relevant to the local population.

Going back to the roots of the problem

The project began as an attempt to stop the ecological degradation taking place around the wells in the pastoralist areas. Its initial efforts were in plantation forestry, which in the mid-1970s was being widely promoted throughout the Sahel. It soon became clear that growing trees for fuelwood or gum production is alien to the pastoralist Fulani or the Wolof farmers. Questions also began to be asked whether tree growing was the answer if over-grazing was the problem.

Going back to the roots of the problem, the project developed a livestock component which came into operation in 1980. The reasoning was that, if the numbers of livestock could be kept at a reasonable level, both animal production and natural regeneration would be improved. To achieve this, areas of land were fenced off and allocated to Fulani families. Initially these had an area of 200 hectares each but new parcels were later extended to 500 hectares. Each was provided with a water point. The number of animals was controlled at predetermined levels by project and government staff.

Livestock production and regeneration of grass and trees were monitored inside and outside the fences. It was found that a range of grasses, herbs and trees were thriving within the fenced plots. The meat and milk production of the enclosed herds was also found to be greater than those kept outside the fence.

One of the key elements in this component of the programme is the decentralisation of water supplies. Provided it is accompanied by measures to manage the livestock and control their numbers, this reduces the heavy local damage from over-grazing and trampling which is such a feature of areas around the large wells.

Project staff are now looking for methods of managing livestock numbers without the use of fencing. Any such system would have to be accompanied by locally accepted control over livestock numbers. As a way of making such controls attractive, the project is now attempting to improve marketing arrangements and set up a livestock transport system under Fulani control. If these efforts succeed, the chances of the programme becoming sustainable, and replicable, will be greatly improved.

The project has thus come a long way since it began in 1975. At that time it was run by foresters and concentrated on pure plantation forestry. This was followed by various forms of imposed village forestry and direct seeding, none of which had much success. Now the project team has an agronomist, a livestock specialist, an economist, a sociologist and an agroforester. This multi-disciplinary team reflects the understanding that the "fight against desertification" is not — as was so widely thought in the 1970s — simply a matter of planting trees.

References.

ANON (1986). "Aufforstung von Brunnenstellen im Norden Senegals".

MUELLER, J.O. Prof. (1986). "Anmerkungen zur Bodenrechts und Gemeindereform in Senegal".

BAUM, U. (1988). "Projektfortschrittsbericht nr.11, 1.6.1988 – 30.11.1988".

MAJJIA VALLEY WINDBREAK PROJECT, Niger

SHELTER FROM THE WIND

The Majjia Valley in Niger acts as a natural wind tunnel, catching and accelerating the strong dry winds which sweep down from the Sahara. Now, after 14 years of tree planting, a seemingly endless series of windbreaks provides a measure of protection to farmers and their fields.

There is no question that the windbreak programme has been a major technical success, it also has reasonable prospects of being economically viable. But there are increasing doubts about the survival of the programme when external funding ends. Questions are being asked about whether enough has been done to involve the local community or indeed whether planting windbreaks is the best way to meet people's needs on a long-term basis. The programme is now diversifying its activities and looking for ways of increasing the involvement and commitment of the local community.

Drought, wind, heat and land degradation

Niger is a large landlocked country with a total area of 1.29 million square kilometres. It is among the hottest and driest places on earth with much of the north and centre covered by the Sahara desert. Temperatures reach 45-50°C during the hot dry season and many areas receive no rain for years. It is also among the world's poorest countries, with an average annual income of US$240 and a life expectancy at birth of just 44 years. In recent decades, the population has been growing at a rate of 2.8% per year.

The only part of the country fit for rainfed agriculture is a narrow strip of land along the southern border with Nigeria. Here, there is a rainy season lasting from June to September and an average annual rainfall of 400-600mm. This area supports about three-quarters of the country's 7 million inhabitants.

It consists of a sedimentary plateau at an altitude of 300-400 metres and is cut by a series of north-south valleys, one of which is Majjia. The plateau itself has few inhabitants and has been heavily over-grazed and deforested. It now

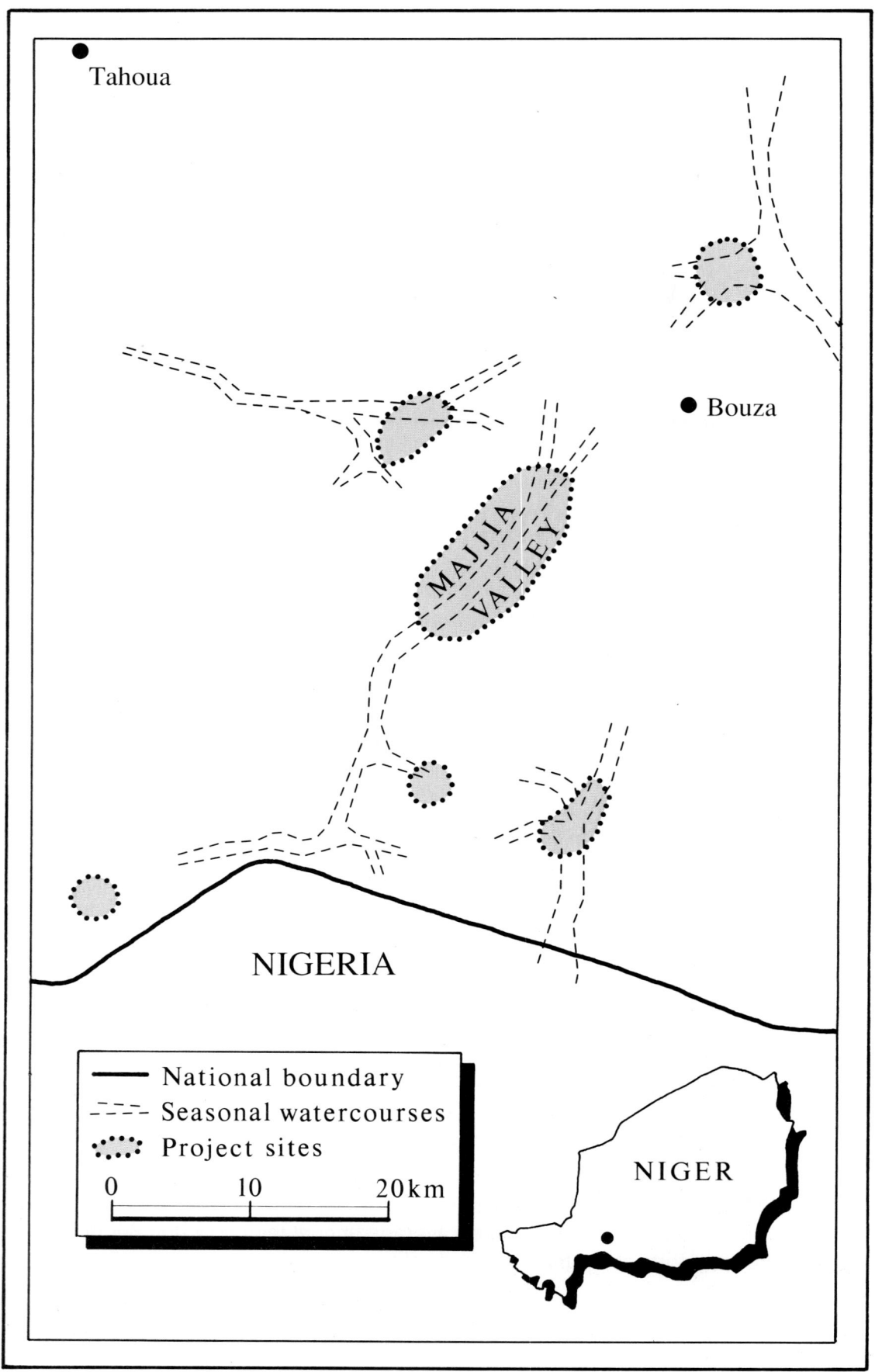

Tahoua

Bouza

MAJJIA VALLEY

NIGERIA

National boundary
Seasonal watercourses
Project sites

0   10   20km

NIGER

Name of Project:	Majjia Valley Windbreak Project (now part of the Management of Semi-arid Lands and Natural Forests Programme)
Address:	c/o CARE Niger, BP 10155 Niamey, Niger. Tel: 740370
Project area:	Five valleys in Tahoea Department, the largest of which is the Majjia Valley
Average rainfall:	400-600mm per year
Implementation:	CARE-Niger and the Forestry Service
Funding:	CARE, USAID, Danida and other donors
1975:	US$40,000
1976-1982:	approximately US$130,000 per year
1983-onwards:	approximately US$300,000 per year
Exchange rate:	300 FCFA = 1 US$ (early 1989)

suffers badly from erosion and land degradation. The vegetation has also been stripped from the valley slopes which too are eroded and unproductive.

The valley floors, however, have deep alluvial soils and high water tables. That in the Majjia Valley, for instance, is at a depth of 4-15 metres throughout the year. The availability of this ground water provides better conditions for agriculture than the average rainfall figures suggest.

Cultivation is intense. Millet is the primary food crop and sorghum is also grown. Cotton, tobacco, water-melon, and tomatoes are other important crops. In the dry season, gardens are cultivated with water from hand-dug wells. Among the tree species are *Acacia albida, Balanites aegyptiaca* and *Ziziphus mauritiana*. Shade and fruit trees are grown in family compounds.

Within the memory of its present inhabitants, the Majjia Valley and the surrounding plateau were rich in forest vegetation and wildlife. Gazelle, lion and other animals were hunted until the environment changed around the 1920s. From then onwards, the natural forest began to degenerate and the process was accelerated by forest clearing and over-grazing.

Life is now extremely difficult for the farming community.

Paul Kerkhof/Panos Pictures

The barren plains surrounding the Majjia Valley were covered in forest until a few generations ago.

Wind erosion is a major problem and farmers may be forced to sow several times because the seeds are buried or blown away. In the rainy season, the valley floors are often inundated by floods which pour down from the valley slopes during rainstorms.

Most of the people living in the Majjia Valley are Hausa farmers while Fulani and Tuareg herders pass through at certain times. As elsewhere in the Sahel, this has given rise to a symbiotic relationship in which the herders provide milk, meat and dung and in return obtain grain and fodder from the farmers. The Tuareg and Fulani may also herd the livestock of the Hausa in return for payment. The Hausa are a matrilinear society: it is the woman who inherits most of the land and she may rent it out to her husband. The Hausa women often keep livestock as an insurance against divorce.

Population growth is putting increasing pressure on the available farming land in the area. The number of people in Majjia is now about 33,000 with a density of 60-75 persons per sqare kilometre. As a result, fallow periods have been shortened and much of the land is under permanent cultivation. There are few local opportunities for earning off-farm income and the majority of the able-bodied men leave Majjia during the dry season to look for work in the cities or abroad.

The windbreak project

In 1974, a local forester conceived the idea of planting windbreaks as a response to complaints by Majjia Valley farmers about wind erosion. The project, which was designed in collaboration with a US Peace Corps volunteer and the international organisation CARE, was launched a year later. Its principal activities were planting neem trees (*Azadirachta indica*) to form windbreaks and stabilisation of a number of dunes. Funding of US$40,000 per year was provided from CARE's own resources.

Initially, the establishment and management of the windbreaks were entirely in the hands of the project. It supplied the seedlings and other inputs required. It planned the layout of the windbreaks, organised and paid the labour through a food-for-work scheme, and generally provided the driving force which has kept the programme going. In recent years, however, the role of the local community has been increased.

The project has also hired the guards required to protect the seedlings until they have grown out of livestock reach. The guards are

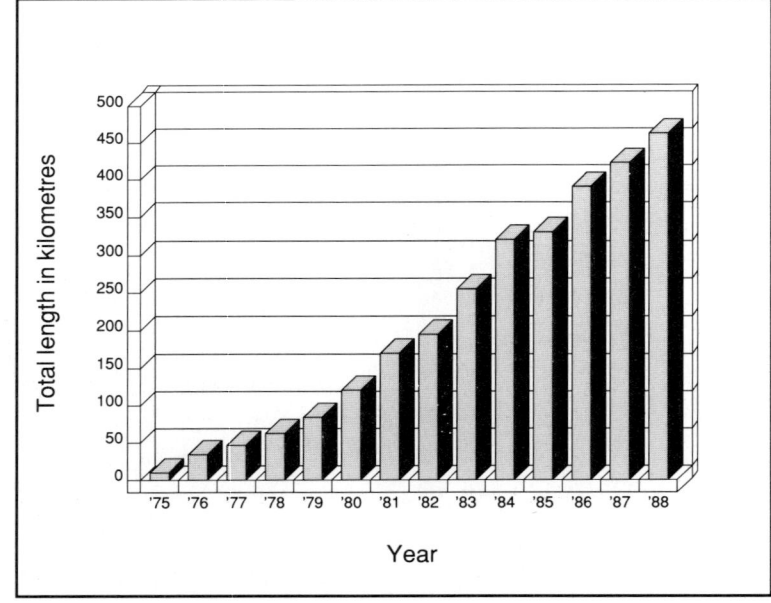

Figure 12.1: Length of windbreaks established — cumulative (1975-88).

Jeffrey Fox/CARE

Windbreaks are spaced 100m apart.   Over 4,600 hectares of farmland are now protected.

supplied with horses which improves their effectiveness.   Livestock found among the newly established windbreaks are impounded and only returned to their owners after payment of a fine.   Initially, fines were about FCFA300 (about US$1) and it turned out that livestock owners were prepared to pay this rather than keep their animals out.   Penalties were later raised to FCFA10,000 (about US$33).

The rate of planting windbreaks has varied from year to year as shown in Figure 12.1.   By the end of 1988, a total of 463 kilometres had been planted and the area of land protected exceeded 4,600 hectares.   The scheme had also been extended to a number of the other valleys in the area.

The project has included other activities besides the establishment of windbreaks.   A number of sand dunes which threatened good agricultural land have been afforested.   It has distributed free seedlings and many trees have been planted in woodlots for pole production, or in compounds for shade and fruits.   An attempt was also made to persuade farmers to grow hedges around vegetable gardens but this was dropped because of a lack of interest.   It appears the farmers dislike the hedges because they occupy valuable land and cannot easily be removed after the cropping season.

Design of the windbreaks

The windbreaks run across the width of the valley and are each about 2 kilometres long.   They consist of a double row of trees at a 4x4 metre spacing, staggered to provide more even wind protection.   This gives a total of about 500 trees per kilometre of windbreak.

The average final height of the trees is about 10 metres.   Since the area protected is about 10 times the height, the windbreaks are spaced at about 100 metres.   In successfully established windbreaks, about 90% of the planted trees reach maturity.

While the Hausa farmers asked for the windbreaks, individual farmers have no say in how the project is implemented. The location and orientation of each windbreak is established regardless of field ownership. Some farmers may have their fields divided by a windbreak, others may; for each farmer, it is simply a matter of luck.

The loss of productive land can be serious. In addition to the 4 metre width of the windbreak, there is a further strip on either side where crop yields are affected by competition with the trees for light and water. With a mature 9-10 year old windbreak, the effective loss in crop land has been estimated at 17%, quite a high figure given the land pressure in the area (Rorison and Dennison, 1987).

Neem is well suited to conditions in the valley and is the main species used in the windbreaks. It is a tall tree which grows quickly. The crown is broad enough to fill the space between the trees but is still permeable enough to let some wind through so that the problem of turbulence, which leads to wind damage of the crops near the trees, is reduced.

Jeffrey Fox/CARE

A well established windbreak consisting of a double row of neem; the project decided where windbreaks should be located, not local farmers.

One problem, however, is that livestock browse the trees up to a height of about two metres. This leads to the creation of a jet of wind between the lower trunks which can cause crop damage and accelerated soil erosion. The more

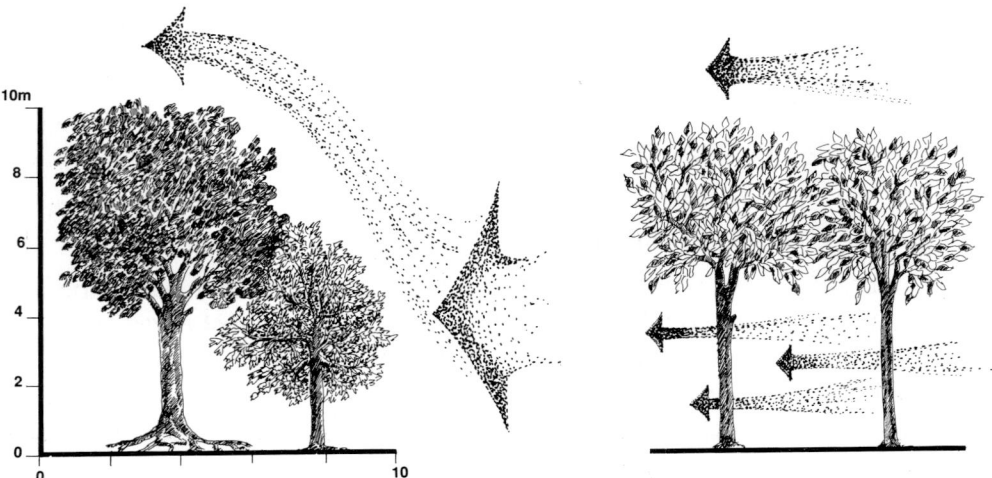

Planting trees of different heights makes a more effective windbreak, according to some researchers.

118

Measuring fuelwood production; part of the extensive research that has been carried out on the impact of the windbreaks.

recent windbreaks therefore consist of a row of neem and a row of the thorny species *Acacia nilotica* which is proof against livestock damage.

The *Acacia nilotica* has so far reached a height of about 7 metres and is planted on the windward side. The wind is thus broken in two steps: first by the lower *Acacia nilotica*, then by the tall neem. In principle, this arrangement provides a better aerodynamic design than the two rows of neem but there is some dispute about whether it really brings much improvement in practice.

Effect on crop yields

The main objective of the Majjia Valley project was to increase crop yields. Windbreaks achieve this by slowing the wind speed, thus cutting down the loss of soil moisture and reducing the mechanical damage to plants.

Studies performed in the valley have demonstrated that the average windspeed in the areas protected by windbreaks has been reduced by 42%. An increase in relative humidity was also recorded (Long and Persaud, 1988). The effect on grain yields is, however, much more difficult to measure. One of the problems is identifying areas of cropland with and without windbreaks which are truly comparable. In areas where improved yields are found, allowance must also be made for the loss of the productive land occupied by the trees and the reduced yields in the crop areas adjacent to the windbreaks.

The early results from the project, however, seemed to confirm that significant net gains were being achieved. A study carried out by a Dutch researcher in 1979, came to the conclusion that yields inside the protected areas were 23% higher than outside, after taking into account the space occupied by the trees

(Bognetteau-Verlinden, 1980). These results received widespread publicity. At last, it seemed as if here was a Sahelian tree growing project which worked.

A much larger study on the impact of the project, called the Majjia Valley Evaluation Study, was carried out in the 1984-85 growing seasons at a cost of US$280,000. This included a sociological survey, wood harvesting studies and an economic analysis. A series of trials were carried out in an attempt to measure the effect of windbreaks on grain yields. These covered over 50 farmers, as opposed to the eight in the earlier study. As 1984 turned out to be a drought year, further research was carried out in 1985.

The results were much less clearcut than in the previous study. In 1984, a 16% increase in grain yields in protected fields, compared to unprotected areas, was recorded. In the following year, however, when two different sets of trials were carried out, the results were not so impressive. One set of trials was tightly controlled by researchers and showed an average 6% increase in grain yield. In the other, which relied more on farmers' control, the results varied widely. In fields protected by nature windbreaks there was no increase in grain yield at all (CARE, 1986). These results were not statistically significant, however, because of the variation between different farmer's fields (Rorison and Dennison, 1987).

The wood harvest

The wood harvested from the windbreaks has turned out to be one of the major benefits of the project. The 1984-85 evaluation study examined a variety of harvesting techniques. All involved pollarding trees at a level above the height reached by browsing animals in order to protect the new shoots. The methods included pollarding every fourth tree in each of the two lines; cutting only the branches projecting over crop fields; pollarding all the trees; and pollarding only the trees in one row of the windbreak.

The studies showed that the method of harvesting can have an important bearing on both the crop and wood yields. The most satisfactory method turned out to be pollarding all the trees in one row of the windbreak. Using this approach, it is estimated that harvesting one kilometer of a 10-year-old windbreak would yield about 900 poles, together with 12 cubic metres of firewood. At local market prices, the poles would be worth about US$1200 and the firewood US$107. Yields from subsequent harvests, assuming a four-year rotation, were expected to be somewhat less than this; 450 poles and 13 cubic metres of firewood, worth a total of US$720.

Sociological aspects

The 1984-85 study also examined some of the sociological aspects of the programme. Three male and three female enumerators interviewed a total of 420 Majjia Valley residents, of whom half were men and half were women. The survey was carried out using a formal questionnaire and was followed up by a small informal survey which dealt with some of the more complicated issues.

The majority of men and women said that, after "lack of rain", wind erosion was the most serious problem for them, with water erosion as a close third. Some 90% of the respondents felt they benefited from the windbreaks. But one of the most telling responses was that only 2% of the villagers thought the windbreaks belonged to them. At that time, the vast majority felt little involvement in the project and believed the trees were owned by the Forestry Department (Delahanty et al, 1985).

The economics of the programme

The 1984-85 study also carried out an economic evaluation of the project. The economic benefits were estimated on the basis of the agronomic and sylvicultural studies. The costs included wages and salaries, material inputs, and the CARE and Forestry Department overheads attributable to the project. A range of interest rates were used.

Benefit calculations were based on the results of the agronomic and wood harvesting trials carried out in 1985. It was assumed that, once windbreaks were established, grain yields would be increased on average by 7.8%, compared to unprotected fields. A four-year harvesting cycle was assumed for the trees, with cutting staggered so that fields would always be partially protected (Dennison, 1986).

The results showed that, under most assumptions, the benefits from the project exceeded its costs. The many uncertainties involved, particularly in estimating the actual increase in crop yields, however, made it difficult to assess the overall economic viability with any precision.

A recent World Bank mission concluded that the project may be having a serious negative impact on livestock production. But this seems debatable given the complexity of the issue. Certainly, the temporary exclusion of livestock from the young windbreaks has a negative impact, especially for the Hausa women; but this has to be set against the longer term increase in the availability of fodder in the form of neem leaves and grain stalks.

The many thousands of neem trees also produce large quantities of neem fruits. Funding has recently been obtained for a study to investigate possible uses for these.

From establishment to management

From the beginning of the project the emphasis has been upon windbreak establishment. In recent years, however, the long-term management and sustainability of the project have been receiving increasing attention.

One of the problems, which was highlighted by the 1984-85 sociological survey, is the lack of local identification with the project. This is primarily because there has been little active engagement by the community or individual farmers in the establishment and management of the windbreaks. The sociological researchers believe that the food-for-work schemes used at various stages of the project have also inhibited the involvement of local people.

This makes it difficult to devise a system which will ensure the long-term management of the windbreaks after the ending of the project. If, for instance, ownership of the windbreaks is handed over to the farmers, they may decide to cut the trees and sell the valuable trunks. If, instead, the Forest Department takes over responsibility, little community involvement can be expected and the same problem of sustainability will arise if funds dry up.

The project has therefore decided that the windbreaks should be communally owned. To achieve this, it has set up a cooperative to manage all the windbreaks in the Majjia Valley. In principle, all adult residents are cooperative members.

One of the key questions is the distribution of the wood products harvested from the windbreaks. Project staff have experimented with a variety of approaches. Under the system presently in use, the major share goes to the cooperative with the rest going to the field owners. The cooperative obtained a total of about FCFA1 million (US$3,300) from the 1988 wood harvest and has begun to invest in a new shop — there are currently no shops in the valley.

The cooperative is, however, still strongly dependent on the project. In one discussion with its leaders, concern was expressed that: "We cannot plant windbreaks, we don't know how to plant in straight lines, and there are farmers who don't want the lines to go through their fields." It is therefore difficult to see new windbreaks being established without outside help; nor is it likely that the present system of protection and management can be sustained by the villagers on their own. The transport of last year's harvest, for instance, relied on project lorries.

This is a matter of considerable concern to the project staff who are working to increase the local community involvement in the programme. Achieving full local sustainability, however, remains a distant goal. As one CARE manager put it: "If we were to leave Majjia now, I wouldn't be surprised if there were no windbreaks left in five years' time." Fortunately, the project has at least another five years to run.

In addition to setting up the cooperative, the project is exploring a variety of other ways of transferring responsibility to local people. A number of farmers, for example, have been trained in nursery management and are raising seedlings under contract. This is seen as a first step towards self-sustaining local seedling production. Digging the holes and planting the trees for new windbreaks is also being left to the farmers concerned in an attempt to increase their involvement.

Beyond windbreaks

The original objective of the Majjia Valley project was to increase food production. Windbreaks were seen as the most effective means of achieving that and the project has been highly successful in establishing them.

The 1984-85 agronomic evaluation has, however, shown that the windbreaks have a moderate effect on crop yields; the main benefit is probably from the production of poles. There is also an emerging worry that the number of trees being planted may lower the water table in the valley. Project staff now accept that windbreaks are not necessarily the only, or even the best, answer to the problems of the Majjia Valley and that the focus of the project has been too narrow.

Research into wind erosion and wind damage to crops elsewhere in Africa has shown that dispersed trees in crop land can be as effective as windbreaks in reducing wind speed. Given that the farmers in the area used to grow various trees, including *Acacia albida*, in their cropland, a programme to encourage this could be an effective approach.

Food security can also be increased by promoting the growing of fruit trees as the project has already done on a small scale. Fodder and fuelwood supplies might be improved by protecting the surrounding slopes and plateau against overgrazing. And, poles can be produced from villagers' own woodlots; many farmers have, in fact, established neem woodlots on their own lands with seedlings provided by the project.

The Government of Niger, CARE, and the donors have recognised this and a much more diverse approach has been planned for the 1988-93 phase of the project which has now been funded. In addition to windbreak establishment, this includes a component to promote intercropping with *Acacia albida* through planting seedlings or encouraging natural regeneration on farmlands. Other components include planting hedges around irrigated plots, woodlots, fruit trees and further decentralisation of seedling production. Training and extension services will also be strengthened.

The Majjia Valley project convincingly demonstrates that a major tree planting project in a semi-arid area can be successful at a technical level. The remaining challenge is to adapt the programme approach to bring about the local involvement and commitment which will ensure that the achievements to date are sustainable and replicable in the longer term.

References:

BOGNETTEAU-VERLINDEN, E. (1980). "Study on the impact of windbreaks in Majjia". CARE-Niger/Agricultural University, Wageningen.

CARE (1986). "Majjia Valley Windbreak Evaluation Niger: Briefing document." CARE.

DELEHANTY, J., M. HOSKINS and J.T. THOMSON (1985). "Majjia Valley Evaluation Study: sociological report". CARE.

DENNISON, S.E. (1986). "Majjia Valley Windbreak Evaluation Study: Economic Evaluation – discussion materials for the Niamey Ad Hoc Committe on Agroforestry." CARE.

DENNISON, S.E. (1988). "The Majjia Valley Windbreak Project Evaluation: A synthesis report." CARE.

LONG, S.P. and N. PERSAUD (1988). "Influence of neem (Azadirachta indica) windbreaks on millet yield, microclimate, and water use in Niger, West Africa". In Proceedings Int. Conf. on Dryland Farming. Amarillo, Texas.

RORISON, K.M. and S.E. DENNISON, (1987). "Majjia Valley Windbreak Evaluation Study: windbreak and windbreak harvesting influences on crop production". CARE.

STEINBERG, D. (1988). "Tree planting for soil conservation. CARE's experience in Niger, West Africa." CARE.

KORO VILLAGE AGROFORESTRY PROJECT, Mali

AGROFORESTRY AMONG THE DOGON

This project was influenced by the Majjia Valley windbreak approach and began with attempts to replicate it among the farmers of the Dogon Plateau in eastern Mali.

The approach originally taken, in which communal windbreaks were imposed on the local people, aroused considerable resentment, however, and progress in the early years was slow. The project has now changed direction and is promoting privately planted windbreaks, decentralised nurseries, tree growing by individual farmers and a variety of soil conservation measures. These are showing promising results and there is considerable optimism about the future.

The area

Koro is a district of 11,000 square kilometres, situated in eastern Mali on the border with Burkina Faso. The climate is Sahelian, with an annual rainfall of 300-600mm. The rainy season is from July to September.

Much of Koro lies to the east of the Bandiagara escarpment and consists of almost completely flat sandy plains. As a result of the increasing population density, shifting cultivation has given way to permanent cultivation in many areas. Millet is the dominant crop and *Acacia albida* and other trees are frequently found intercropped in the fields. There is little cash crop cultivation and monetary incomes are generally low. During the dry season, male migration to the towns in search of work is common.

The majority of the population, about 78%, are Dogon farmers; the next largest group, about 12%, are Fulani cattle herders. As elsewhere in the Sahel, the farmers have a symbiotic relationship with the herders. The herders provide animal products and may also look after the livestock of the farmers; the Dogon, in their turn, supply millet and other crops and allow grazing in their fields after the harvest.

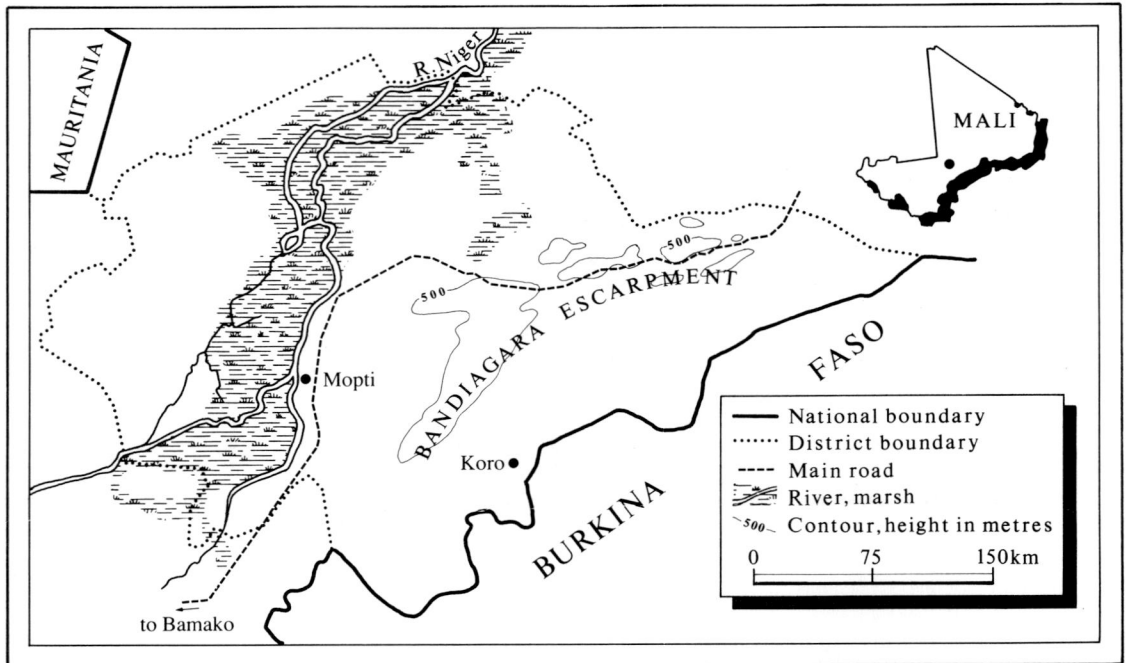

Name of project:	Koro Village Agroforestry Project
Address:	CARE-Mali, BP 1766 Bamako, Mali. Tel: 222262
Project area:	Cercle de Koro (Koro District), Mali
Average rainfall:	Annual average 300-600 mm
Implementation:	CARE-Mali with the Forest Service
Funding:	CARE, USAID, NORAD
1983-86:	US$200,000 per year
1986-90:	US$400,000 per year
Exchange rate:	300 FCFA = 1 US$ (early 1989)

What differentiates the Dogon from many other ethnic groups is their culture and religion. Despite the extremely strong influence of Islam in the area, they have retained many of their animistic religious traditions and rituals. They are also renowned for the quality of their wood carvings.

The photogenic villages, and unusual celebrations and rites, have made the Dogon country into a major tourist attraction. Art traders from all over the world can also be found scavenging the area in the hope of finding some of the remaining older carvings which have now become almost priceless.

The Koro Village Agroforestry Project

The Koro Agricultural Extension Project was begun in 1983. Initially it was funded by CARE; additional support was later provided by USAID and more recently by NORAD. The funding was US$200,000 per year during the period 1983-86 and was increased to about US$400,000 per year for the period

126

1986-90. The implementing agency is CARE-Mali together with the Forest Service.

The area suffered badly during the droughts of the 1970s and early 1980s. Wind erosion and declining soil fertility were among the other problems facing the farming community. At the time the project was being planned, it was felt that windbreaks would reduce the problem of wind erosion and increase agricultural yields; the research results then emerging from the Majjia Valley project suggested that yields might be increased by around 20%.

The project also believed that obtaining domestic fuel supplies was a major problem. This view, which was widespread at the time, was based on anecdotal evidence such as, for example, a report that "in some cases women in the 5th region of Mali walk 20 kilometres one way to gather fuelwood." It was therefore assumed that the windbreaks would satisfy a significant part of local fuelwood needs.

In 1985, an excursion, for project staff, administrators and farmers, to view the windbreaks in Majjia Valley was organised by CARE. The visit impressed the participants to such an extent that windbreaks were seen as the answer for Koro. The project manager tried, in vain, to dampen this enthusiasm.

The same windbreak design as in the Majjia Valley was used. The trees were planted in two rows at a staggered spacing of 4x4 metres perpendicular to the predominant wind direction. The early results were not encouraging. In 1985, for instance, 2 kilometres of windbreaks were planted but by June 1986 only 12% of the trees had survived.

Despite this, the Forest Service and local councillors insisted on continuing with the project, relying, if necessary, on strong-arm methods. In 1986, a total of 19 kilometres of windbreaks were established, in 1987 about 30 kilometres. Much of this planting, however, succumbed to fire, drought, browsing, or the plough.

One of the reasons for this was purely technical. While the water table in the Majjia Valley is generally less than 15 metres below the surface, in Koro it is normally beyond the reach of tree roots. Thus seedlings were more at risk from drought.

More important, however, was the coercive approach which village authorities and the Forest Service used to get the windbreaks established. In some cases, villagers were forced to establish "communal" windbreaks on chiefs' fields. While these tactics could ensure that seedlings were planted, they provided no motivation for protecting them and hence the generally poor survival rates.

Changing the direction of the project

By 1987, it was clear that a new strategy was needed and from late in that year the project began to change its approach and diversify its activities. The emphasis was placed on initiatives which the local people felt were relevant; low-cost solutions were encouraged rather than the expensive windbreaks; and the emphasis was shifted towards individual rather than communal actions.

Windbreaks are still being established, but they are now being planted mainly by private individuals who choose to do so, on their own land. The principal species are *Acacia raddiana* and *Azadirachta indica*, and the newly planted trees are protected with thorn branches or millet stalks. In 1988, a total of 43 kilometres of windbreaks were planted and the survival rate of the seedlings after one year was 80%.

The pattern of planting in private windbreaks, of course, bears no resemblance to that used in the Majjia Valley and whether the windbreaks are going to be effective in increasing food production has yet to be demonstrated. In fact, there are some indications that the main interest on the part of the farmers is in using the trees for wood rather than protecting their lands from the wind.

Communal windbreaks have mostly failed, but some farmers have planted windbreaks on their own land; this farmer has used millet stalks to protect his seedlings.

The project is also endeavouring to promote the natural regeneration of *Acacia albida* on farm lands. The area is typical *Acacia albida* country and the Dogon have traditionally been well-aware of the virtues of this unique species. In recent times, however, interest in encouraging its regeneration seems to have diminished, especially among the young.

The increasing land pressure is perhaps one of the reasons for this. Another is that cutting trees or branches without a permit is forbidden by law and this deters farmers from letting trees grow on their lands. In practice, the majority of seedlings which have naturally regenerated are deliberately ploughed under or browsed by animals.

The project has subjected some 80 fields, covering a total of 120 hectares, to a number of different treatments to encourage regeneration. In some cases, goat dung, which tends to contain a considerable number of *Acacia albida* seeds, has been applied; in others, the tree seeds have been sown. The young seedlings are marked with a stick or some other means to protect them against the plough.

The result has been an abundant regeneration. This, however, can only happen in a period when the rains are good, as they have been in the past few years. But it does clearly demonstrates what can be done when the circumstances are favourable. Moreover, the Dogon seem to prefer such low-cost and low-risk methods of regeneration to tree planting.

Since the Fulani herders cut *Acacia albida* branches during the dry season, their cooperation in such regeneration efforts is obviously essential. Despite its efforts to do so, the project has not yet managed to involve them in these activities.

The project is also promoting other soil conservation techniques on the Seno plain. One of these is the "Zay", a type of microcatchment which is mulched to achieve greater water penetration. The model is derived from Yatenga Province in Burkina Faso (see PAF project profile).

In another initiative in 1988, some villages established rock bunds on about 3 hectares of communal land. The resulting increase in millet production encouraged the villages to extend the work to another 30 hectares in 1989. The project is also promoting the use of live fences and tree planting along field boundary lines.

All this amounts to a major change from the Majjia Valley windbreak model on which the project was based during 1985-88. It has led not just to a more broadly based project but to one which is much more closely tuned to local needs. In effect, it is a recognition that tree growing alone is rarely an answer to the problems faced by rural people in areas such as Koro.

Developing a sustainable method of seedling production and distribution

Before the project began there was only one dilapidated Forest Service nursery in existence in Koro. This was rehabilitated by the project in 1983 and since then three more central nurseries have been established. During the period 1983-87, a wide variety of seedlings were produced in these nurseries. The species included *Eucalyptus camaldulensis*, *Azadirachta indica*, *Acacia raddiana*, *Acacia albida*, *Prosopis juliflora*, *Prosopis chilensis*, and fruit trees such as guava, papaya and baobab.

Paul Kerkhoff/Panos Pictures

A heavily fenced nursery; the project is now encouraging decentralised seedling production.

The four nurseries were run more or less directly by CARE staff until 1988. Since then, CARE has attempted to increase the involvement of the Forest Service and each nursery now has a nursery head and assistant from the Service. The project still provides one additional full-time worker for each nursery and some temporary staff during the peak season.

Initially the seedlings were given free of charge to anyone asking for them. Experience showed that people did not take much care of free seedlings and there is now a charge of FCFA50 (16 US cents) each for forestry trees and FCFA200 (65 UScents) for fruit trees. When seedlings are used for specific project activities, such as the establishment of windbreaks, farmers do not have to pay for them.

The costs of seedling production and distribution using central nurseries are, however, high. There is little chance the Forest Service would be able to bear them after the end of the project. If the project is to be sustainable, it is imperative that a cheaper decentralised system is established.

In order to encourage such decentralisation, the project has stepped up the promotion of micro-nurseries owned by individual farmers. This has so far been relatively successful as can be seen in Table 13.1, where the outputs from the central and micro-nurseries are compared. Production from the micro-nurseries was more or less equal to that from the central nurseries in 1988 and exceeded it by 50% in 1989 (Mansius, 1989b).

The nursery owners are usually men but there were nine women-operated nurseries in 1989. The owners are initially trained by project extension staff; they are also provided with free polythene bags, seeds and tools. Most of the seedlings are bought by the project at a price of FCFA50 per seedling (16 US cents). This is cheaper than growing them in one of the central nurseries.

The micro-nurseries use water and take up space close to the village well. This is tolerated by the villagers, perhaps because they now have a nearby source of seedlings. The fact that the price of the seedlings is lower than in Koro town also helps make the micro-nurseries acceptable.

The aim is to create a self-sustaining seedling production system at a village level. The free inputs and the high level of dependence on sales to the project give some cause for worry on that score. Nevertheless, a certain proportion of the trees, particularly fruit species like papaya, are being sold to local people. Such sales may well continue in the absence of the project. A number of nursery owners have said that they would be prepared to pay for the plastic bags if they were not provided free. The project management has decided to reduce the project inputs further in order to reinforce the tendency towards self-sustainability.

Antipathy towards the Forest Service

The Forest Service in Koro has provided eight agents who work half-time on project activities. Although the Forest Service personnel perform a certain number of extension tasks, the bulk of the extension workers have been hired by CARE, and are local Dogon people.

In recent years an increasing number of women extension agents have been appointed and the proportion is now five women to four men. The high proportion of women in the team is a deliberate project

Table 13.1

Annual seedling production in central nurseries and micro-nurseries

	central nurseries	micro-nurseries
before 1988 (average)	60,000	few
1988	52,000	50,000
1989	52,000	76,000

policy. All are equipped with mopeds, an effective and popular form of transport in Mali.

The reason for not using Forest Service staff for extension is primarily because of the deep local antipathy felt towards them. This is because the Forest Service has maintained the colonial tradition of policing the rural population and has even carried this into non-forestry areas such as improved stoves. It is the Forest Service which fines people for not having an improved stove or for not using it.

Farmers are also fined for felling trees or cutting branches, even if the trees are situated on their own fields. Permits for cutting can be obtained, but it may take a farmer a day or two to walk to and from the nearest Forest Service station. These fines are collected by the Forest Service staff; and sometimes they are imposed in great numbers. They can cause severe hardship to local people as illustrated by an example from the 5th Region, which is unconnected to the project.

"That morning Amadou Diallo sent his son to collect firewood. Amadou himself set out to help a neighbour, while the 10-year-old Mussa rode quite a distance with the ox-cart to an area where the villagers normally collect firewood. Later in the morning, when Amadou was still helping his neighbour, the bad news spread. While young Mussa had been collecting firewood, he was caught by a forester. Both oxen and cart were taken. Amadou was shattered. He knew he had lost his cart and his only ploughing ox, and that he would be unable to raise the money to pay the fine. The work was stopped for the rest of the day."

Foresters are legally entitled to keep a proportion of all the fines they levy for themselves; many are thought to keep the whole fine. Given that their wages are extremely low, and are often paid 3-6 months late, their revenue from fines is essential to their economic survival. It can also account for the enthusiasm with which this aspect of their work is carried out.

In 1983, the foresters in Koro District, some of whom are, in fact, ex-policemen, collected FCFA1.3 million in fines (US$4,300); by 1988 this figure had increased FCFA14.9 million ($47,000). Forest Service officials sometimes mobilise motor-cycle units which move rapidly from village to village fining villagers who are found breaking the rules. The natural result is that the Forest Service has a poor local image. It is well aware of this, but given the economic strait-jacket in which it finds itself, radical change is not easy.

Looking to the future

The project has moved a long way since it began in 1983. Rather than the compulsory planting of communal windbreaks, it now relies on fostering a variety of voluntary actions by individual farmers.

But it has also become evident that even if responsibilities can be devolved right down to a village level, tree planting in Koro remains a difficult operation. If the other avenues of action currently being explored, such as the promotion of natural regeneration, direct seeding, and soil conservation, prove to be successful, tree planting will become less necessary. This has been adopted as the long-term objective of the project.

This approach, however, depends on having skilled extension workers who are acceptable to the local population. It is extremely difficult to see the Forest Service effectively taking on this role. If the project is to have a long-term future, it may be necessary to shift the responsibility for running it to another Ministry.

References:

BENAFEL, D. (1982). "Survey of Koro Cercle Villages". Memorandum. CARE.

CAMPBELL, P. (1985). "Village Agroforestry Project – FY'86". CARE.

HAGEN, R. et al (1986). "Koro Village Agroforestry Project, an evaluation report". CARE.

MANSIUS, D. (1989a). "Koro Village Agroforestry Project, Annual Implementation Plan, FY 1989." CARE.

MANSIUS, D. (1989b). Personal communication.

PROJET AGRO-FORESTIER, Burkina Faso

INCREASED CROP YIELDS ARE FIRST PRIORITY

After several years of effort, this small agroforestry project abandoned its efforts to promote tree growing. It found that farmers were much more concerned about their crop yields than growing trees.

It then introduced a water-harvesting method based on the use of stone contour bunds which succeeded in raising agricultural production. This innovation has been widely adopted and is spontaneously spreading outside the project area. Now that crop yields have improved, the project is finding that farmers have become more responsive to suggestions that they should plant trees and improve natural regeneration.

Problems in Yatenga Province

The average rainfall in Yatenga Province in the north of Burkina Faso declined in the 1970s and 1980s from an annual average of over 600mm to 400-500mm. Yet in spite of these worsening conditions, the population has continued to grow. According to the 1985 census, the population density in the area was 40 persons per square kilometre which is very high for a semi-arid region; in some parts, the density is even higher.

The majority of the people in the area are settled farmers, but there are also some semi-nomadic pastoralists. Because of the increasing population density, much of the natural savanna forest and bushland has been cleared and converted to agricultural use. This has resulted in severe land degradation in places since many of the soils are heavy and form a hard crust when exposed to the sun. The problems have been exacerbated by over-grazing in some areas.

Production of food crops is often far below family needs. Many people, especially the young men, leave during the dry season in search of temporary employment in Côte d'Ivoire.

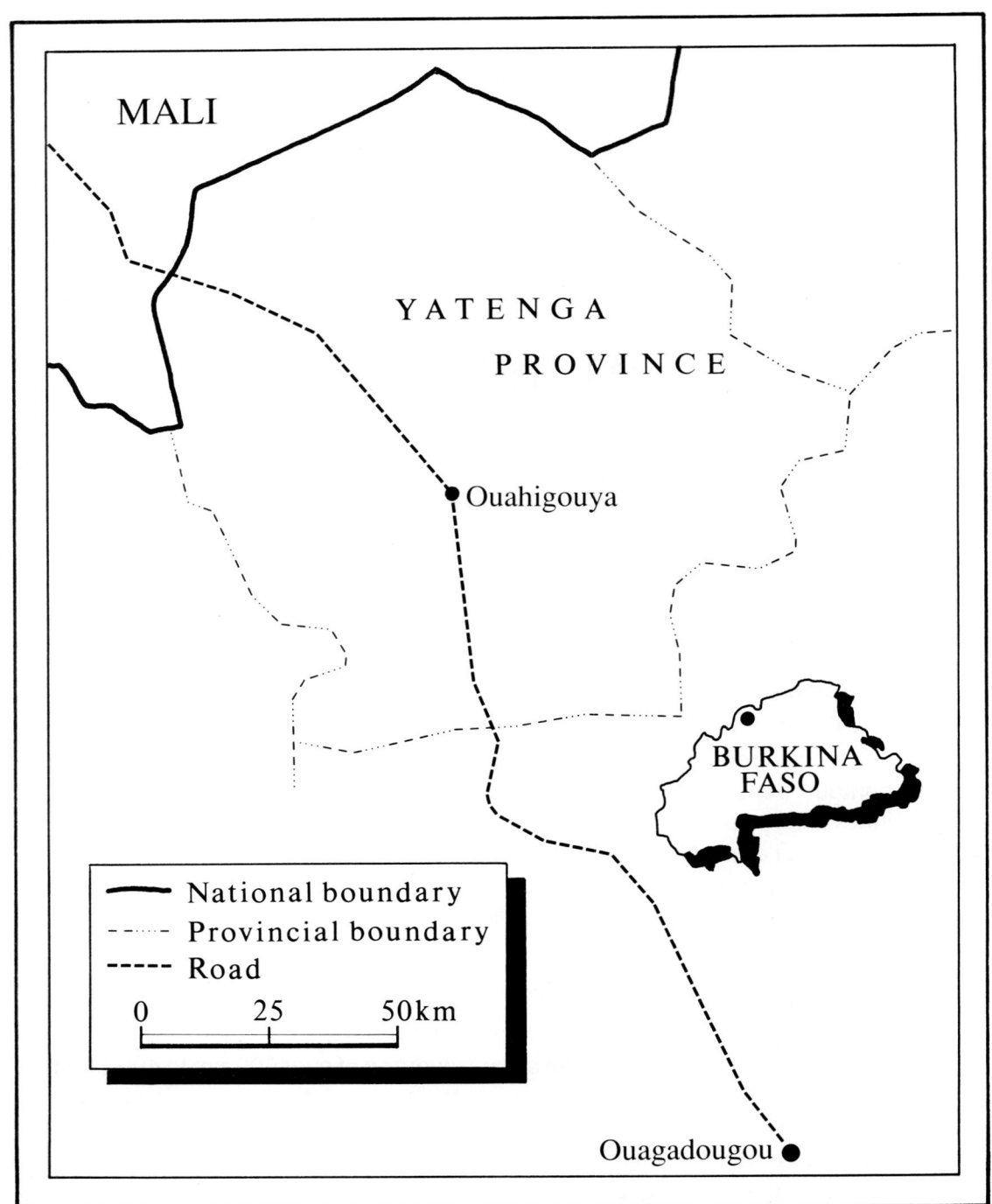

Name of project:	**Projet Agro-forestier (PAF)**
Address:	**c/o Oxfam, B.P. 489, Oagadougou, Burkina Faso.**
Project area:	**Yatenga Province**
Average rainfall:	**400-600 mm per year**
Implementation:	**Oxfam with the Ministry of Agriculture**
Funding:	**Oxfam - U.K.**
1979-87:	**US$116,000 excluding expatriate costs**
1987-89:	**US$162,000 excluding expatriate costs**
Exchange Rate:	**300 FCFA = 1 US$ (early 1989)**

Lack of interest in tree growing

The project began in 1979 by promoting the use of microcatchments to enable farmers to grow trees for wood. The technique was similar to that used in the Negev desert in Israel. It was felt that trees would be better than food crops in utilising the water harvested by the microcatchments.

The early years were seen as experimental and a variety of techniques were used to construct the microcatchments. In 1979, a total of 200 microcatchments covering a few hectares were installed and there were trial plantings of some 20 different indigenous and exotic tree species. In 1981, the area treated was 7 hectares and in 1982 it was 62 hectares. A total of 4,100 seedlings were produced in 1982 from eight nurseries.

From the beginning, however, the farmers showed little enthusiasm for tree planting. The 1982 annual report of the project noted that "stronger extension is necessary to make sure farmers grow trees among their crops". Seedling survival rates were also poor. Goats were identified as a major cause of damage and project staff started fencing the trees but this was a costly measure.

It became apparent that the root cause of the problem was that tree growing was not a priority for farmers. They were in favour of the microcatchments because they helped increase food production but they had little interest in the trees.

Fortunately, the project had the flexibility to change its direction in response to this realisation. In 1982, the tree component was shelved in favour of promoting rock bunds as a means of increasing crop production.

Using rock bunds to increase crop production

Initially, the project had experimented with the use of earth bunds to form the microcatchments. The main advantage of these is that the necessary material is available in the fields. The problem is that they need a great deal of maintenance which many farmers, already over-burdened with work at critical periods in the farming year, are unable to manage.

The emphasis was therefore shifted to the use of rock bunds. These have long been used by the Dogon farmers in Mali who build them in straight lines across the slope. The project innovation was to introduce the idea of building them

Jeremy Hartley/Oxfam

Rock bunds are laid out at intervals across the field.

135

along the contours. This is done using a water level made of a hosepipe and two sticks.

Rock bunds require a much higher initial investment because of the arduous task of collecting and transporting the necessary rocks. Farmers may have to travel several kilometres to find a source of suitable material. But once the bund has been established, there is little maintenance required.

Figure 14.1 A cross-section of a rock bund.

The design finally chosen by the project embeds the bund about 10 cm into the ground to prevent it being eroded by water flowing beneath it. The height is typically 20-30cm above the ground. The rocks used may be as small as gravel or up to the size of a football (see Figure 14.1). The actual line of the bunds is determined by the soil and whether there are gullies, patches of bare rock or other features. The spacing between bunds depends on the gradient but is normally 10-15 metres.

In addition to the rock bunds, the project is also promoting a number of other improved agricultural practices. One of these is the "Zay" method of tillage. This is a traditional practice whereby a 20x20cm hole with a depth of 10cm is dug during the dry season and filled with mulch such as crop residue. This leads to increased termite activity which in turn increases the rate of water penetration when the rains come. Millet is planted in the individual Zay holes which also help to protect the seedlings from wind damage. The project also

Mark Edwards/Oxfam

Building a rock bund; water levels are used to find the contours.

promotes the use of an improved compost heap as part of a national programme to improve composting by farmers.

Since 1983, the rock bunds, the Zay and the improved composting form the project package instead of tree planting. But PAF has retained its original designation as an "agroforestry project".

Effective local participation

Rather than relying on food-for-work as an incentive, as is commonly done in the region, PAF decided that the rock bunds should be built by villagers themselves without payment. The project was, however, prepared to assist in providing information and training. It also supplied materials and equipment such as water-levels, pick-axes, and carts either free or on loan. Some food loans were also given, most of which were paid back after the following harvest.

Despite the lack of incentives, the rock bunds have been widely adopted. Local people are convinced they repay the input of time and labour by providing higher yields and increasing the security of the harvest. The bunds also extend the area under cultivation by enabling land which had previously been considered useless to be brought into production.

Some older men mentioned their reason for wanting the rock bunds: "So many young people have left ... if we cannot improve life in our village, and all young people leave, who will look after us when we grow very old?"

Villagers collecting rocks on a hill several kilometres from their fields. Some of the tools are provided by the project.

Project staff have carried out measurements comparing grain yields on plots with and without rock bunds over a number of years. These showed that the average yields in the plots with bunds were consistently higher than in those without bunds, ranging from 12% in 1982 to 91% in 1984. Although such results may not be significant in a statistical sense, they are amply borne out by field observations. At times the difference between treated and untreated fields is quite spectacular.

The increased food production is obviously of crucial importance to the Yatenga farmers. Rock bunds are now found throughout the province even where they have not been directly promoted by PAF or any other project. The technique is also spreading to other parts of the country and many excursions are made from other provinces to see the progress in "soil and water conservation in Yatenga". By the end of 1988, the total area of land treated with rock bunds under the programme was estimated to be about 2,600 hectares. It is estimated that a further 3,500 hectares had been covered by farmers who had independently adopted the technique. In all, the method has been adopted by about 500 villages.

Paul Kerkhof/Panos Pictures

Making the most of the available extension resources

The project has been building up its extension programme since 1983. Initially, it relied on the use of the flannel board and other techniques to raise awareness. It then moved on to providing training in the construction of rock bunds.

Since it is too costly to provide individual instruction to all farmers wishing to construct bunds, PAF asked villages to nominate representatives who could learn the technique and then pass on the knowledge to others. In all, several thousand people have now been trained, covering about 500 villages.

The problem is that PAF is a small organisation with limited resources. In 1989, the total extension staff was five people, four men and one woman. The Ministry of Agriculture, on the other hand, is a large organisation with many extension agents in Yatenga Province. It is, however, extremely short of operational resources like transport, tools and communication equipment.

A formal agreement on collaboration in extension work was therefore made with the Ministry. Under this, PAF pays the Ministry's extension staff a fixed sum of FCFA4,500 (US$15) for petrol and other costs while they work part-time on the project. They provide instruction in the building of bunds as well as supporting the extension work of villagers who have undergone the training programme. The number of Ministry staff involved has increased from 31 in 1985 to 81 in 1989.

Rock bunds have proved to be highly effective in increasing crop yields in their immediate vicinity.

Initially, the focus of the training and extension work was on men farmers but this has been changing steadily. During the 1985-86 dry season, the project began training women farmers in rock bund construction and women are now included in the extension team.

After food security back to growing trees

Project staff have found that attitudes to tree growing tend to change once the project package has been implemented. In some villages, farmers are now prepared to plant trees along the bunds if they are provided with seedlings by the Forestry Department; equally importantly, they are also willing to provide protection against grazing animals during the dry season.

This is partly because they feel that the problem of food security, which is their primary concern, has been addressed. It is also because the construction of the bunds creates improved conditions for seedling survival. Grasses and perennials are able to benefit from the increased soil moisture and grow among

Mark Edwards/Oxfam

A neem tree planted in front of a bund; once bunds are established farmers seem to be more open to the idea of tree planting.

the rocks of the bunds. As a result, young trees and shrubs are able to survive the dry season provided they are not too heavily browsed.

Recko village, for instance, attempted to establish a village woodlot as part of the national woodlot programme some years ago, but without success. Recently, a variety of trees were planted along the bunds which had been established on about 5 hectares of agricultural land. In this case, the better conditions created by the bunds and the commitment to keep animals out of the upgraded fields has made tree growing much more practicable.

In fact, the ease with which natural regeneration takes place under these conditions may mean that it turns out to be more important than tree planting. This is something the villagers are interested in finding out for themselves.

In Longa village, the use of bunds has been integrated into the activities of another project which is concentrating on land use management. Here, the villagers constructed rock bunds on 70 hectares of agricultural land. They also agreed to impose severe restrictions on the movement of animals

All livestock is now held in shaded enclosures and fodder is collected from the cultivated areas. Trees have been widely planted and look excellent, as does the natural regeneration which is shooting up. Neighbouring villages are now copying the example of Longa. Seedlings are supplied by the Forestry Department and guidance on stall feeding is provided by the Ministry of Agriculture's extension agents. The costs of stall construction and other extras are borne by the villagers themselves.

Stall feeding under the dry conditions of the area is an unusual initiative but several other villages have already expressed their intention of copying Longa. It remains to be seen whether the controls hold out under the stress of drought. If they do, it could be the beginning of another exciting development.

Lessons from success

One of the principal lessons of the project is the importance of mobilising the community. Many development projects have switched the emphasis to individual, or family-based, activities when they have run into difficulties in promoting communal action. In 1985, PAF itself noted that "farmers are interested in individual rock bund construction". It also found that the quality of rock bunds on individual land was better than those on communal land.

In Longa village the community has decided that all their animals should be stall-fed. This has made tree growing easier and has facilitated natural regeneration.

As time passed, however, it became clear that much of the work can only be done effectively if there is a community consensus. If, for example, rocks are not available nearby, their collection needs to be organised on a community basis; if land is to be protected from grazing, the animals of all farmers without exception must be controlled. The communal orientation, if it works, also helps to ensure that the poorer farmers are included among the project beneficiaries.

The level of community organisation, however, varies greatly between villages. In some cases, the project has entered into collaboration with the section of the village which has agreed to cooperate on rock bund construction. The remaining farmers have often followed suit once they have seen the work in progress. In other cases, the project simply waits until the village has organised itself.

Because of the progress which has been made, there is no longer any need for measures to raise the general level of awareness. People see the innovations in neighbouring villages and discuss them as part of their ordinary day to day contacts.

In areas where the innovations have not yet been adopted, the project promotes exchange and communication through excursions. The villagers are then left to reflect on what they have seen, discuss the issues, and organise themselves when they feel ready for action. The farmers know where to find the project office in Ouahigouya. In the final analysis it is up to them to get together and undertake development activities for their own benefit.

The project also recognises that problems remain. So far, less than 10% of the farmers have constructed rock bunds on their lands. There are villages where, despite the evident success of the technique in some fields, adjacent farmers have not yet adopted it. A considerable amount therefore remains to be learned about the attitudes and constraints which operate at a local level.

There is also the question of uneven distribution of costs and benefits. Rock collection and construction of the bunds make heavy demands on the available labour, and probably most of all, on women. Rich farmers are also more likely to be able to mobilise and provide food for communal groups to build bunds on their land. Indeed, poor farmers may owe work to the richer farmers which makes it even more difficult for them to construct bunds for themselves.

The sustainability of the increased yields obtained when using the bunds is another cause of concern. Higher crop production means greater mineral extraction from the land. There is therefore a danger of long-term soil depletion unless methods of increasing the input of organic matter and fertilisers can be put into operation. Thus the recovery of manure from stall-fed animals and the encouragement of composting assume a critical importance in the longer-term perspective.

The project accepts the need to identify and resolve such issues, but is fully aware that whatever solution or technical package it develops must be acceptable to the local population. Project staff take pains to point out that the reasons for its success are not primarily technical. The techniques it has developed work because they fit their local context and meet the farmers' need for low-risk and low-investment strategies. More important than its ability to develop technical packages is the fact that the project has won the confidence of the local people and that they accept that it is concerned with their needs and priorities.

Thus there is no attempt to persuade villagers to do anything they do not believe in or want. When it was clear that tree growing was not a local priority, it was abandoned. The project is prepared to wait until the people themselves are ready to take the initiative and it then supports and helps them. This demands an open-minded, competent and dedicated project team. If such a team can be put together, as in this case, it adds up to a resoundingly successful project strategy.

References:

MATHIEU, O. (1986). "Rapport d'activités de la campagne 1985". Oxfam, Burkina Faso.

OXFAM (1979-83). "Annual Reports, PAF/Micro-catchments in Upper Volta."

WRIGHT, P.L. (1988). "Rapport mission sur orientations du projet agro-forestier du 9 au 29 Novembre 1988". Oxfam.

SOIL CONSERVATION AND AGROFORESTRY PROJECT, Zambia

SEEKING AN APPROPRIATE PACKAGE

Declining agricultural yields and recurrent drought are among the major problems faced by the Tonga people in southern Zambia. Widespread crop failures have brought famine to some areas in recent years. Attempts to increase food production by means of labour intensive soil conservation measures were strongly resisted by local farmers. A proposal to introduce ecological farming techniques also proved unacceptable because of the complex intercropping patterns required. It now appears that a tree growing programme run by a local NGO is leading to a balanced and effective approach.

The project area

The project covers an area of about 520 square kilometres in the Mazabuku region of southern Zambia, some 100 kilometres to the south-west of Lusaka. It is at an altitude of about 1,000 metres and the topography is generally flat. The climate is sub-humid, almost semi-arid, with an average annual rainfall of 780mm. Like most of Zambia, and in comparison with much of the rest of Africa, the area is quite well covered with trees.

The Tonga are a farming people with a substantial number of livestock. Maize is the main subsistence food crop while sweet potatoes, cow-peas, sorghum and several vegetables are also cultivated. Maize and the recently introduced cotton are the major cash crops. *Acacia albida* is a traditional part of the farming landscape and these vast trees are widely dispersed in the croplands. Miombo woodland covers much of the area not under cultivation.

Part of the project area was designated as trust lands in colonial times and land tenure is subject to traditional communal customs. Here, the population density is 65 persons per square kilometre which is high compared with the rest of the country. The rest of the project area, which was formerly occupied by white farmers, is a resettlement scheme with a population density of 30 persons per square kilometre.

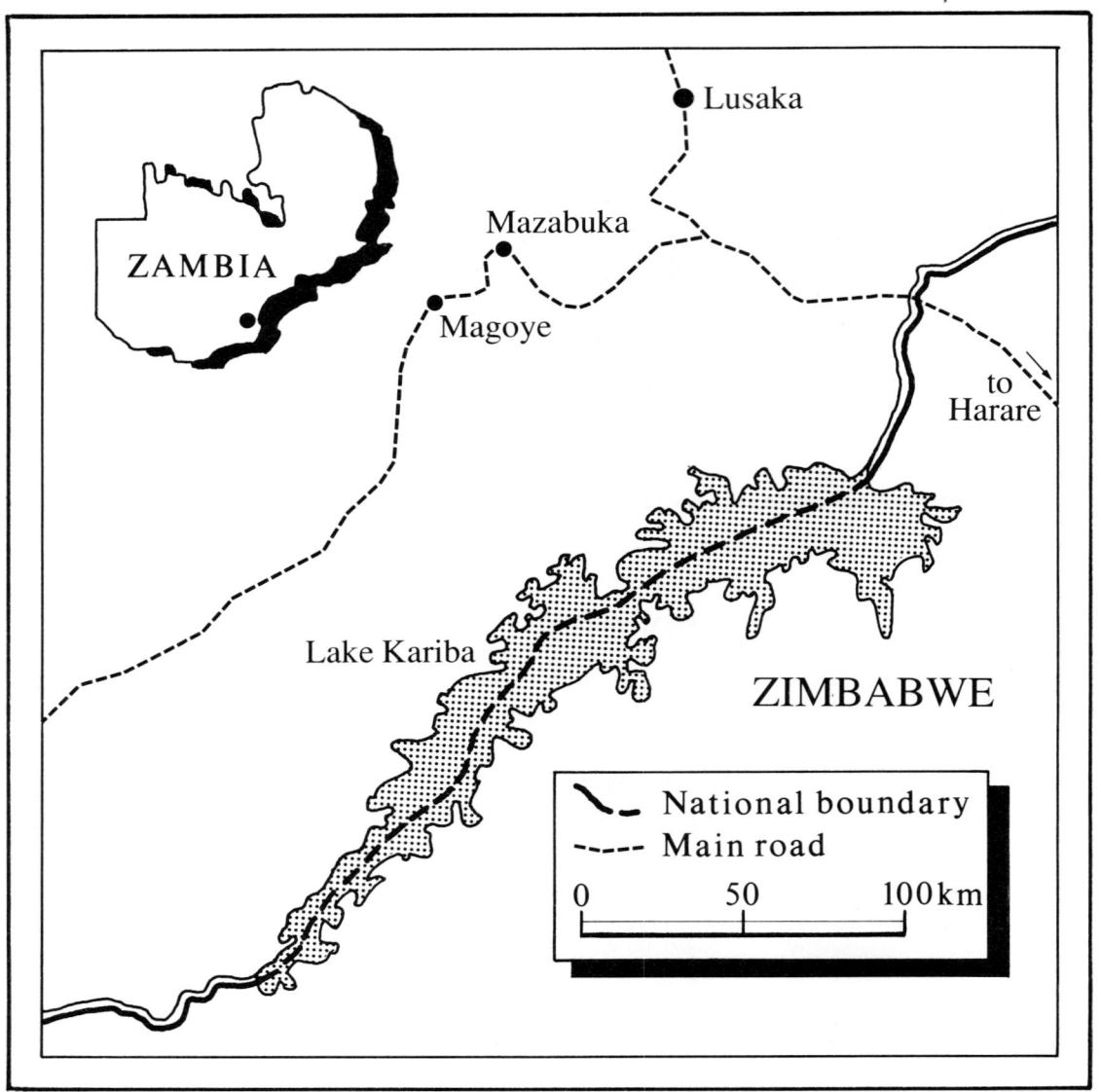

Name of project:	**Soil Conservation and Agroforestry Project**
Address:	**P.O.Box 42, Magoye, Zambia.**
Project area:	**Lusume area in Magoye District**
Average rainfall:	**780 mm per year**
Implementation:	**Lusume Services, a local NGO.**
Funding:	**General support from Government of Zambia, United Church of Canada, German Volunteer Service. Specific grant of US$14,800 from German Agro Action.**
Exchange rate:	**17 Kwacha = 1 US$ (October 1989)**

*Large **Acacia albida** trees are found among many farmers' fields; but there is little natural regeneration.*

Early approaches

Lusume Services is an NGO which is active in several rural development activities and receives general assistance from the United Church of Canada. It has its headquarters in Magoye in the north of the project area. In 1984, it established a soil and water conservation unit with support from the Government of Zambia and the German Volunteer Service.

The initial programme relied purely on mechanical soil conservation measures. It concentrated on activities such as building dams, promoting contour ploughing and the construction of contour bunds. This was highly unpopular among farmers since it involved them in a great deal of work, especially at the ploughing and harvest seasons when the demands on their time are already at a maximum. It also reminded them of the forced labour that was imposed during colonial times.

The emphasis was therefore switched to soil conservation through improved agricultural practices, particularly the use of permanent ground cover. Ideas about ecological farming were gained from an international course held in The Netherlands. Stimulated by this, the team established experiments near the project headquarters in 1985. These included the use of cover crops, nitrogen fixing crops, crop rotation and natural pesticides. But it soon became clear that the complexity of the systems involved would not be acceptable to local farmers. The dense spacing of leguminous trees, for example, would prevent oxen-weeding. The majority of the experiments were therefore abandoned.

The present programme

The present programme began in 1986, when a decision was made to promote intercropping with *Acacia albida* as a means of increasing the ground cover and improving the fertility of the soil. Other project objectives included the promotion of tree planting in windbreaks and on field boundaries; the use of live fences; planting trees and shrubs on contours; raising awareness about impending fuelwood shortages; encouraging the planting of fruit trees; and growing trees for fodder. A grant of US$14,800 was given to the project by German Agro Action.

The promotion of *Acacia albida* was a new policy in the area. Although the farmers are familiar with the species and its benefits, the number of trees has been declining in recent years. This can be partly attributed to the Department of Agriculture which has a well-staffed extension service in the area and

145

discourages intercropping of trees in farmland. In the eyes of the Department of Agriculture, a "progressive" farmer is one who has removed all the trees from his cropland.

The project also decided to establish its own nurseries since the Forestry Department only provided a selection of traditional forest species such as eucalyptus. A total of about 10,000 seedlings, mainly *Acacia albida*, were produced in 1986. Fruit tree seedlings were also produced in smaller numbers for sale to the public.

Women do much of the work on the farm, but rarely take part in project meetings; a special extension scheme has been started to try and involve women in the project.

In the same year, as a means of spreading the project message, staff officers participated in farmers' meetings organised by the local cotton company. During these, time was spent discussing agroforestry and, particularly, the benefits of intercropping with *Acacia albida*. Some 40 farmers showed an interest and were formally registered with the project. During the rainy season, they were each provided with 25 seedlings free of charge. The project specified that these should be planted in croplands at a regular spacing of 12x12 metres.

This specification was based mainly on research on *Acacia albida* in West Africa. It turned out to be too rigid for the farmers, and they planted the trees in a variety in configurations. They also made it known to the project that they wanted a wider selection of species. The 1987 nursery production therefore included a broader range, including some fruit trees, and the output was raised to 20,000 seedlings.

During the dry season, audio-visual presentations of the agricultural problems in the area and the proposed project package were made at farmers' meetings. Leaflets about conservation and intercropping with *Acacia albida* were also produced in the local language, CiTonga. As a result of these efforts, some 100 farmers were registered for follow-up with seedlings during the rains; this number has continued to increase.

The proportion of women participants in the meetings was, however, only about 5%. This was a serious weakness in the programme since women do most of the agricultural work. Project staff have come across several instances in which men had planted *Acacia albida* seedlings which were subsequently pulled up by their wives who considered them to be weeds.

Separate meetings are now held, mostly in schools, which reach a large

146

number of women and children. The project has also recognised that women extension officers are more likely to have an impact on the women and young men working on the farms.

Problems with research

A new set of trials was begun in 1986, at the Lusume Services headquarters in Magoye. Instead of the previous ecological farming trials, these concentrated on multipurpose trees and alley cropping.

A total of 64 tree species and varieties were planted on a 1 hectare plot using a variety of spacings and intercropping systems. They included nitrogen fixing shrubs like *Leucaena leucocephala* and *Sesbania sesban*; hedge species like *Ceasalpinia decapetala* and *Ziziphus abyssinica;* fruit trees such as mango and papaya; and trees with pesticidal properties like *Azadirachta indica* and *Melia azadirachta.*

The trials were essential to obtain information on species performance since there has been very little agroforestry research in Zambia. They were also intended to give project staff practical experience in agroforestry as well as acting as a demonstration for local farmers.

Initially the experimental lay-out was simple, but in 1986 the ICRAF recommendations for carrying out such trials were adopted. These are designed to provide the necessary information for a detailed statistical analysis of the results, but they add considerably to the cost and complexity of the trials. By 1988, a certain amount of data on wood yields had been obtained but little useful information was derived from the intercropping trials since some of the crop yields had been mixed up.

Nevertheless, the trials have produced some valuable, if negative, results, *Leucaena leucocephela*, still considered a miracle species in much of Africa, shows a survival rate of less than 20% in the project area. This is mainly because it is highly prone to termite attack. Where *leucaena* is planted on farms it is also browsed and destroyed by livestock.

The alley cropping system has also turned out to be impractical for inclusion in the programme. Preliminary results from the trials indicate that 5,000-6,000 trees per hectare are needed if a significant contribution is to be made to soil fertility and wood production. This is highly labour intensive and there is thus little possibility of it being adopted by local farmers on a widespread basis.

A belated diagnostic survey

A survey to assess the agroforestry needs of the area was not carried out until 1987, because of time pressure on the programme. In the beginning, it was felt that the available resources had to be concentrated on creating a visible impact in the field and this feeling of urgency led to the postponement of the survey.

When it was carried out, the survey revealed that tree planting is common in the project area, but is almost entirely confined to fruit trees such as mango, citrus, mulberry, papaya, and guava. The popular places for planting trees are around the homesteads and as windbreaks, but not among agricultural crops. The survey also revealed that, while *Acacia albida* is appreciated by most farmers, it is not planted (Sturmheit et al, 1988).

These results explain why so few farmers were interested in taking the *Acacia albida* seedlings. If the information had been available earlier it would undoubtedly have saved a certain amount of time and resources.

Useful gains and an uncertain project future

It is now clear that planting *Acacia albida* in croplands is not particularly popular among farmers. This is mainly because of the labour involved and the fact that the planted seedlings require a considerable amount of protection against fire and livestock. Neither are farmers prepared to plant trees for fuelwood or fodder on their croplands, although they might be more willing to try it on grazing lands.

The project can nevertheless show a number of useful gains. Because of the publicity, farmers increasingly recognise *Acacia albida* as a valuable tree and instead of weeding them out are leaving seedlings growing, and a programme to promote this has been incorporated in the project. There has also been a certain amount of success in encouraging the planting of windbreaks and live fences, and some farmers appear to be starting their own small nurseries.

All this has been achieved with very modest means. Nevertheless, the long-term problem facing the project, if funding comes to an end, is that some of its key elements are not self-sustaining. The tree nursery, for instance, is only manageable with paid labour and as long as transport is available at the onset of the rainy season. Project staff recognise this: "We must promote establishment of small nurseries in the individual camps and villages." The question is how to do it.

It is obviously difficult for a small NGO to change long-entrenched official attitudes such as those in the Department of Agriculture and the Forestry Department. Nevertheless the project has already shown that something can be done to change farmers' attitudes to *Acacia albida.* It has demonstrated that farmers are responsive to tree planting if appropriate species are available. It has shown that women can be reached in agricultural extension, if given due attention.

If such lessons were to be taken up by the Department of Agriculture with its large and permanent extension service, it could have a significant effect on agricultural development in Zambia.

References:

STURMHEIT, P. (1988), "Traditional agricultural practice as a model for famine prevention: Acacia albida extension in Zambia's Southern Province".

STURMHEIT, P. et al (1988). "Evaluation of a soil conservation and agroforestry needs assessment study conducted in Mazabuka District of Zambia."

THE HADO PROJECT, Tanzania

EVICTING THE CATTLE

Land degradation on a spectacular scale has long been a feature of the Dodoma Region of central Tanzania. Major efforts were made to halt the destruction by means of grazing controls and soil protection measures but these proved inadequate to deal with the magnitude of the problem. The government therefore took the drastic step of completely evicting the cattle from one of the worst affected areas.

The improvement in the vegetation cover has been dramatic and the land is now far less vulnerable to erosion. Despite initial opposition, a high proportion of the local people are now in favour of the scheme. But there have also been social costs and other special features of the project which need to be taken into account when it is being considered as a model for replication elsewhere.

A history of soil erosion

The project area is in the Kondoa District which lies to the north of the Dodoma Region in central Tanzania. The climate is hot and semi-arid with an annual rainfall of 600-800mm.

The majority of people earn their living by farming but there is also a tradition of pastoralism in the area. Even farming families tend to keep cattle, sometimes in substantial numbers. Livestock are regarded as an insurance against crop failure in the harsh and uncertain climate. The Rangi people are the dominant ethnic group.

The soils are light and sandy and there is a long history of over-grazing and soil degradation. The heavy rainstorms which come at the end of the dry season are a particularly potent cause of erosion; up to 40 mm can fall in a couple of hours. Travellers in the last century reported "dust devils" and "denuded hills".

During colonial times efforts were made to force the population to adopt a variety of anti-erosion measures such as contour banking, ridge cultivation and reduction of livestock numbers. These were highly unpopular and were

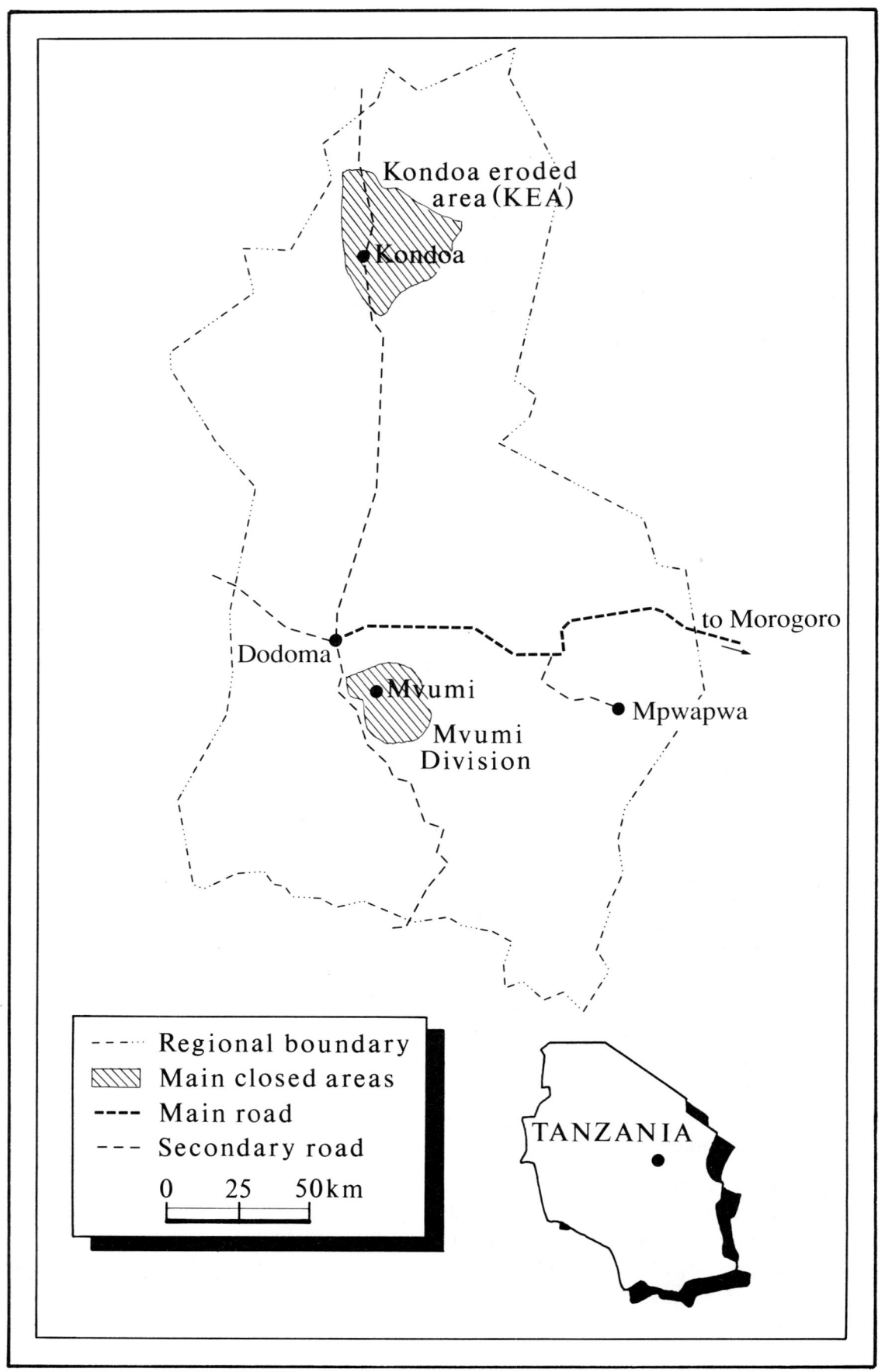

Regional boundary
Main closed areas
Main road
Secondary road

0   25   50km

Kondoa eroded area(KEA)
Kondoa
to Morogoro
Dodoma
Mvumi
Mvumi Division
Mpwapwa

TANZANIA

Name of project:	Hifadhi Ardhi Dodoma (HADO)
Address:	HADO, P.O.Box 840, Dodoma, Tanzania.
Project area:	Dodoma Region
Average rainfall:	600-800 mm per year
Implementation:	Forestry Department
Funding:1973-83:	total US$0.8 million
	1986-96: total US$1.0 million
	SIDA provides about 75% of the funding
Exchange rate:	120 TSh = 1 US$ (late 1988)

abandoned after independence (Mbegu and Mlenge, 1983). In the years immediately following independence, however, soil erosion became extremely severe.

In 1968, the Local District Council, on the advice of the Central Government, enacted a bye-law prohibiting grazing, cultivating, digging water channels for irrigation or cutting trees within the area without permission. This had little practical impact and the land destruction continued virtually unabated.

Severe gully erosion is common in the Kondoa Eroded Area; in 1979, livestock was completely banned from the area.

HADO established

When President Nyerere visited the area in 1972 he was shocked by the erosion taking place and ordered the Ministry of Natural Resources to take immediate action. A team from the Division of Forestry was appointed to investigate the issue and recommended a national soil conservation project with headquarters in Kondoa. This was established in 1973 as a special project directly responsible to the Director of Forestry in the Ministry of Natural Resources and Tourism. The official title is the Ardhi Hifadhi Dodoma (HADO) project; the name means "soil conservation in Dodoma". HADO's activities are not confined to forestry; its formal brief is to conserve soil and water and rehabilitate degraded land.

Although HADO's responsibilities cover the whole of the Dodoma region it has concentrated its activities in the Irangi Highlands in Kondoa District. This is an area where the land degradation and gulley erosion are famous for their severity. It is officially designated as the Kondoa Eroded Area (KEA). It covers 1,256 square kilometres and has a population of about 74,000.

151

HADO began its work with a vigorous programme of tree growing and soil conservation. Each year, from 1973 to 1979, several hundred hectares of demonstration woodlots were established and around 400,000 seedlings were distributed to villages, schools and other institutions.

The soil conservation measures included construction of contour bunds on slopes, erection of earth or stone check dams in gulleys, and the use of quick sprouting vegetation along stream banks to control flooding and erosion. By 1979, the area treated in this way was about 7,300 hectares.

At that stage, HADO, reviewing its progress, calculated that it would take 105 years to cover the rest of the KEA. Given that the question was whether the area would be habitable in another 10-20 years, it was clear that a different strategy was required.

The move to total destocking

At that stage, the density of livestock in the KEA was 69 animals per square kilometre or a total of some 90,000 animals. As these were clearly the main cause of the land degradation, the need to reduce their numbers substantially was obvious. The question was how to do it; the precedents were not encouraging.

One suggestion was that a strict limit be set on the number of livestock which a family could own. This was considered to be impractical because of the difficulty in enforcing it and because families would share their livestock among relatives. Faced with the urgency of the problem and the absence of feasible alternatives, HADO proposed the total destocking of the area.

No one was under any illusions about the sensitivity or complexity of the issue. The proposal was discussed and approved at a meeting of the District Executive Committee of TANU, the national ruling party, in April 1979. It was agreed that all cattle, goats, sheep, mules and donkeys were to be removed from the KEA by the end of October, thus giving farmers six months notice.

An intensive one-week campaign was carried out by HADO together with local Party and administrative staff to inform people of the measure. Senior Party representatives and national politicians, including the Minister for Natural Resources and Tourism, also toured the area and addressed meetings in support of the decision. Each village was given a map, a boundary description of the KEA, and a list of suggested villages outside the area which, because of their low livestock density, could safely absorb additional animals.

Although owners of substantial numbers of livestock in the KEA were in a minority, making up about 25% of the population, they were generally among the more powerful members of the local community. They strongly opposed the destocking and vowed to campaign against any of the elected community leaders or Members of Parliament who supported the destocking. They also argued that the move was depriving people, not just livestock owners, of milk, meat and manure. HADO's basic answer was that, in the absence of such a measure, the area would be unable to support any livestock within a decade or two.

When the destocking time came, some livestock owners did not believe it would actually happen and hid their animals. The policy was, however, firmly implemented and an estimated total of some 46,000 cattle, 29,000 goats and 11,000 sheep were removed from the area.

No compensation was provided to the families involved. Some decided to move their homes and established themselves in the neighbouring lands. Others sent their livestock away with family members.

It was not an easy time for anyone concerned. A local government officer serving in one of the affected villages said: "All were in despair those days. We who supported the decision found ourselves all of a sudden being the enemies of the people. It was right to expel the livestock. But when the day came when the decision had to be be enacted, it was very very difficult" (Östberg, 1986). There were, however, many who accepted the decision.

During the ten years since the destocking, the regulations have continued to be rigorously enforced. Exceptions are only made for draught oxen during the ploughing season, donkeys used for pulling carts, and animals brought in for slaughter. The HADO staff are responsible for policing the area, supported by the local administration.

People convicted of breaking the regulations face fines of up 1000 TSh (about US$8). HADO staff, who are unarmed, often find it difficult to catch trespassers or obtain convictions when they bring offenders to court. One HADO forester was killed by a trespassing cattle herder in 1983. The number of court cases brought by HADO during the period 1979-84 is shown in Figure 16.1; an average of about 60% of these resulted in convictions.

Changing tree growing strategies

In addition to its destocking activities, HADO has continued to promote tree growing. Seedling production in the period 1979-83 was an average of 1.7 million per year. These were used for demonstration woodlots, soil conservation measures, and decoration and avenue planting. They were also distributed free to villages, schools and institutions.

A total of about 2,600 hectares of demonstration woodlots had been established by 1983. In a number of places, farmers, who were angry that the woodlots had been established on land which they believed they were entitled to cultivate, retaliated by burning them. There have also been problems of livestock damage and weed competition. The more successful woodlots have, nevertheless, been producing wood for several years and this is sold by HADO.

Creating woodlots is, however, expensive and HADO is now concentrating primarily on

TABLE 16.1. Number of cases in two primary courts of KEA where people were accused by HADO of illegal grazing

Year	Number of court cases
1979	45
1980	51
1981	30
1982	56
1983	29
1984	37

Source: Oestberg, 1980

distributing seedlings to schools, villages and individual farmers. Dodoma Rural District, for instance, has four nurseries from which seedlings are transported to the villages. The local village administration is normally responsible for distributing the seedlings to the individual villagers. Decentralisation of seedling production to villages and schools is considered important but little headway in achieving this has been made so far.

The regrowth of grass and woody biomass has been dramatic in protected area.

The range of species in the nurseries has been widened in the light of experience. There is an increased number of fruit trees available and more recently, leguminous species have been promoted. Multi-purpose trees like *Leucaena leucocephala* have also been given more attention.

Few villages have created communal woodlots; where they have been established they have tended to be neglected. But in densely settled areas tree planting is popular. In some places there is a local market for timber, and poles and firewood can be sold in Kondoa town. Families increasingly regard having a few trees growing in the compound as a form of insurance. They can be sold if there is a need for cash; in this respect they fill the same role as cattle.

HADO has also obtained the help of Sokoine University at Morogoro in the design of a number of agroforestry trials. Some are testing the intercropping of cereals with *leucaena*. Others are looking at the production of fodder for stall-feeding cattle. None of these has, however, been running long enough to produce any firm data.

Results of livestock eviction

It is clear that the destocking has resulted in a number of major changes in the area. The most noticeable is the transformation of the landscape which now has a heavy grass cover throughout the year.

The increase in vegetation cover has meant that run-off and erosion have been drastically decreased all over the KEA. This has led to a reduction in floods with the result that rivers have become narrower and remain in flow for a longer period. The amount of cultivation in or near river courses has consequently increased considerably.

In one instance an area of 200 hectares of sand has been stabilised and is now used for crop growing. Sweet potato, for example, is one of the crops now frequently cultivated. It is used for consumption at home, for barter against millet from the plains, or is sold for cash.

Crop cultivation has also increased in the low-lying areas or "mbuga". Some

Paul Kerkhof/Panos Pictures

With better vegetation cover, the flow in rivers is now more steady and crops like sweet potato can be cultivated in riverbeds.

of these are marshes in which it is possible to grow crops even in the dry season. These have traditionally been used as dry season grazing reserves but now, because of the absence of cattle, can be used by farmers.

The cropland available after destocking has been evenly distributed between farmers whether or not they formerly owned cattle. Thus a certain redistribution of wealth has taken place as a result of the project. It has also been reported that fuelwood supplies have increased because of the greater bush productivity.

Overall, there has been a considerable increase in agricultural output with benefits in cash and nutrition for farmers. Previously the KEA was one of the areas of the country which had obtained continuous food relief supplies. A study in 1986 reported that no emergency food supplies were needed in the previous three years although they had been provided for the surrounding areas (Östberg, 1986).

The eviction of the livestock has, however, also had negative effects. Lack of milk is a recurrent theme in discussions with local farmers on the impact of the project. One of its uses was to produce the fat for cooking vegetables. People now use sunflower oil as a substitute and a small cottage industry in its extraction seems to be developing.

Farmers also miss the availability of dung for manure and have to rely instead on chemical fertilisers. These are expensive and at times are not available; in the 1981-82 season, for example, there were no supplies. Some farmers are turning to mulching as a substitute.

The improved vegetation cover is also causing some unpredicted problems. The great increase in dry biomass allows fierce fires to sweep through the area causing considerable damage. These are often caused by farmers burning their fields, or by honey gatherers or hunters, and are extremely difficult to prevent or control once they have started. The availability of food and hiding places has also led to a large increase in wildlife, with the numbers of warthogs, monkeys, snakes, dik-dik and birds all increasing. In some cases, these are causing severe crop losses.

Bullrush millet, for example, is an extremely popular food crop in the area but needs to be guarded carefully against quela quela birds. Because of the difficulty of protecting the more distant fields, its cultivation has had to be abandoned in some areas (Östberg, 1986). There have also been suggestions that the incidence of malaria has increased. In some of the areas surrounding the KEA, to which the livestock were moved, there have been reports of over-grazing and a certain amount of erosion.

The overall balance sheet is, however, clearly positive. The apparently inexorable environmental decline of the area has been halted and agricultural production is rising. Young people who would previously have emigrated are apparently staying in the area. Some of those who left with their cattle have returned feeling that, despite the lack of livestock, the area offers them a better livelihood. Although some former cattle owners still remain bitter about what has happened, the majority of people, including former cattle owners, approve of what has been done, or at least accept its necessity.

Paul Kerkhoff/Panos Pictures

Lining up at a health clinic in Mvumi Division; the livestock ban has had a big impact on agriculture, nutrition and health, but little is known about these effects.

Replicating the HADO model

HADO's work in the KEA represents a breakthrough in dealing with the previously intractable problem of severe land erosion caused by over-grazing. It is highly regarded by government administrators and TANU officials and in recent years it has been used as a model for replication elsewhere.

One such area is the Mvumi Division in Dodoma Rural District. This has a total area of 713 square kilometres and is seriously affected by over-grazing and soil erosion. In 1986, it was completely destocked; the total number of cattle evicted was 65,000.

Although the basic KEA model was used, a number of changes were made and the scheme was more thoroughly prepared. An initial survey was carried out in Mvumi and the surrounding areas to determine livestock movements. In addition, more effort was put into informing the people of what was going to happen and why.

The publicity campaign lasted about a year. Meetings were held in all 13 villages to educate the village government in soil conservation and destocking. Seminars were held for progressive farmers, local technical staff and administrators. As a result, the destocking went more smoothly than in Kondoa.

HADO also introduced a stall-feeding project for cattle. At present, three farmers are running stall-feeding units and the project has established fodder plots which are used for seed and grass production. A veterinary team has been deployed by a local NGO. HADO has also brought an improved bull to Mvumi. Given the small number of farmers with zero grazing units, the impact of this project component is still limited. What is important is that it shows HADO is taking on wider responsibilities.

A series of complementary soil conservation measures were also carried out in the hills of Mvumi. Much of this work, such as digging cut-off drains which divert storm floods away from the farm land, is done by the villagers themselves. The Tanzanian Prime Minister, Joseph Warioba, spent three days helping in this work, which gave HADO additional political leverage.

Another total destocking project is at Guyikrum Hill in Karatu in northern Tanzania. Although nobody was living in the area, it was extensively used for grazing and was showing signs of land degradation similar to Kondoa. In 1985, the Forestry Department together with the District Council introduced a programme for the complete removal of livestock from an area of about 1000 ha. All grazing, cutting and burning were also prohibited.

In the first year, there were numerous prosecutions but within a few years the regulations were being respected. Once the hill had been protected it recovered very quickly and is now covered with dense vegetation.

Since the area concerned is much smaller than KEA and Mvumi, nobody is really hard hit by the protection measures. In addition to restoring the hill, the project is benefiting the local farmers who have been given access to the area to cut grass which they sell for thatching or fodder.

The same types of initiatives have been taken in other areas suffering from severe over-grazing and erosion. They are usually on a small scale and do not always take the form of a project with special funds. They may simply be a response by local foresters and councillors to protect a badly eroded hill in their area.

Broadening the approach

One of the noticeable features of the programme HADO programme is that, in spite of its multi-sectorial impact, the professional staff of HADO has, up to very recently, consisted entirely of foresters. The project recognises the need to broaden its approach and has recently employed a land-use specialist.

But it can be difficult to coordinate the field activities of different ministries concerned with rural developmental, particularly at the higher levels. A forester and a livestock extension worker in a village may be convinced that close cooperation is necessary but, as a project staff member said, "there is little we can do if the directors in Dar es Salaam do not feel the same way."

One measure being taken is to increase the amount of training for project staff and local people. In the first phase of the project, up to 1979, the

Paul Kerkhof/Panos Pictures

Stall feeding is being encouraged in Mvumi Division to offset the effects of the livestock ban.

amount provided was minimal. In the second phase, it has been greatly increased. The training provided in 1987, for example, was as follows:

* 25 HADO senior staff	5 days (each)
* 50 Agriculture and livestock staff	5 days
* 84 Local councillors	5 days
* 21 School teachers	3 days
* 133 Farmers	3 days

It is recognised that much more training is required at all levels in the administration and the community itself if there is to be an adequate understanding of how to develop and implement effective soil conservation techniques once destocking has taken place. In particular, improved communication is required between HADO and the local people to reduce the almost inevitable tensions which arise when people are compelled to remove their livestock from their grazing areas.

There is also a need for an evolution in attitudes and policies as circumstances change. The prohibition on cutting trees and keeping goats makes sense in the beginning when natural regeneration is just starting. As the bush recovers, however, means of controlling it and restricting the wildlife it harbours have to be developed.

The issue of when and under what conditions the livestock can be readmitted must also be faced. Some people feel, for example, that since oxen are allowed into the area for ploughing, no harm would be caused if they stayed permanently. The answers to such evolving questions will have to be worked out in collaboration with the local community if the gains already achieved are not to be put at risk.

Pointers to the future

Deforestation and land degradation in arid and semi-arid lands are on the increase all over Africa. Tree planting and mechanical soil conservation measures have largely failed because they do not address the real problem, which is over-grazing. HADO is one of the first official organisations to put its full weight behind a policy of drastic destocking, despite the political hazards involved.

The benefits in the KEA are clear but the particular circumstances of the project must also be borne in mind when it is being used as a model. One of the most important factors contributing to its success was that there were alternative lands available nearby which could accommodate the livestock and those families which did not want to give up their herds.

Another important point was that the Rangi people are skilled and diligent farmers. They have been able to take advantage of the opportunities for increased agricultural production which were created by the project. A purely pastoralist people would be unlikely to respond so positively to the loss of their livestock.

There is also a need to ensure that destocking projects are supported by an adequate spread of professional disciplines. It is to the credit of the HADO foresters that they were able to achieve so much. But as the KEA experience suggests, the regeneration of the natural vegetation in an area may well bring unexpected problems such as fires, wild animals and a resurgence of insect pests. Helping farmers adjust to rapidly changing ecological conditions and deal with such problems requires a level of expertise beyond that normally expected from foresters.

The way in which destocking projects are implemented will also vary depending on local circumstances. The initial HADO approach has, with some justification, been described as "high handed" and "authoritarian" (Leach and Mearns, 1988). The time limit for getting rid of the cattle was short and people were faced with a non-negotiable administrative decision. It is significant that in subsequent destocking exercises greater efforts have been made to win the support of local people in advance.

The more subtle approach used in Giyukrum Hill may, indeed, prove to be more effective in the long term. This has the attraction of changing and adapting the livestock economy rather than breaking it up entirely. The people who live in the immediate vicinity are also able to enjoy the right to sustainable exploitation of the area quite soon after the protection is effective.

The HADO experience confirms what has been found elsewhere in Africa. Tree planting is rarely, if ever, an effective response to the problem of land degradation in arid or semi-arid zones. Nor can erosion control measures be implemented at the scale and rapidity required to deal with the problems of land degradation which are plaguing so many of these regions.

There is now no doubt that severe destocking is required if the complete collapse of many of these areas is to be prevented in the next couple of decades. It is to the credit of HADO that it faced the difficult choice and acted decisively and effectively. The project has shown what can be done given the necessary political commitment and effective enforcement measures. The challenge is to learn from this experience and develop a range of approaches which can be applied where social and political conditions are different.

References:

HADO (1986). "HADO Project Phase Two Project Master Plan: 1986/87 – 1995/96". Ministry of Natural Resources and Tourism.

LEACH, G. and MEARNS, R. (1988) "Beyond the woodfuel crisis: people land and trees in Africa." Earthscan Publications, London.

MBEGU, A.C. and W.C. MLENGE (undated) "Ten years of HADO: 1973- 1983". Forestry Division, Ministry of Natural Resources and Tourism.

OSTBERG, W. (1986). "The Kondoa transformation". Scandinavian Institute for African Studies, Research report No. 76.

TURKANA RURAL DEVELOPMENT PROJECT, Kenya

INVOLVING PASTORALISTS IN MANAGEMENT OF FORESTS

Early attempts to promote tree growing among the fiercely independent herdsmen of the Turkana region showed little success. The project has now shifted its emphasis to participatory management of the natural vegetation and is showing much better results.

The Turkana Region

Turkana, in northern Kenya, is the driest area in East Africa. It borders Uganda, Ethiopia and Sudan and covers an area of 72,000 square kilometres. The average rainfall varies from 400 mm down to 180 mm per year, far below that necessary for settled rain-fed agriculture. All the rivers in the area are seasonal and the major ones are lined by extensive forests.

The area is inhabited by the Turkana, a pastoral people with a rich traditional knowledge of their environment. Their life is a complex blend of nomadism and settled farming which varies depending on circumstances. Many settle on a temporary basis along the seasonal rivers. Once the rains are over, some family members, together with most of the livestock, leave for areas where grass can be found. The remainder, mainly older people and some of the women and children, stay behind with the rest of the stock and may grow some crops. But in some cases, the whole family moves.

While most of the land in the district is considered communal, some of the forest areas along the rivers are divided into plots, called "ekwar", which are traditionally owned by individual families. These plots produce fruits, construction material, medicine and most importantly, browse for livestock.

The dominant tree species in the riverine forests are *Acacia tortilis* and *Acacia eliator*. In the dry season, the Turkana beat these trees with sticks, up to 10 metres long, to knock down the seed pods to feed their animals. Another

161

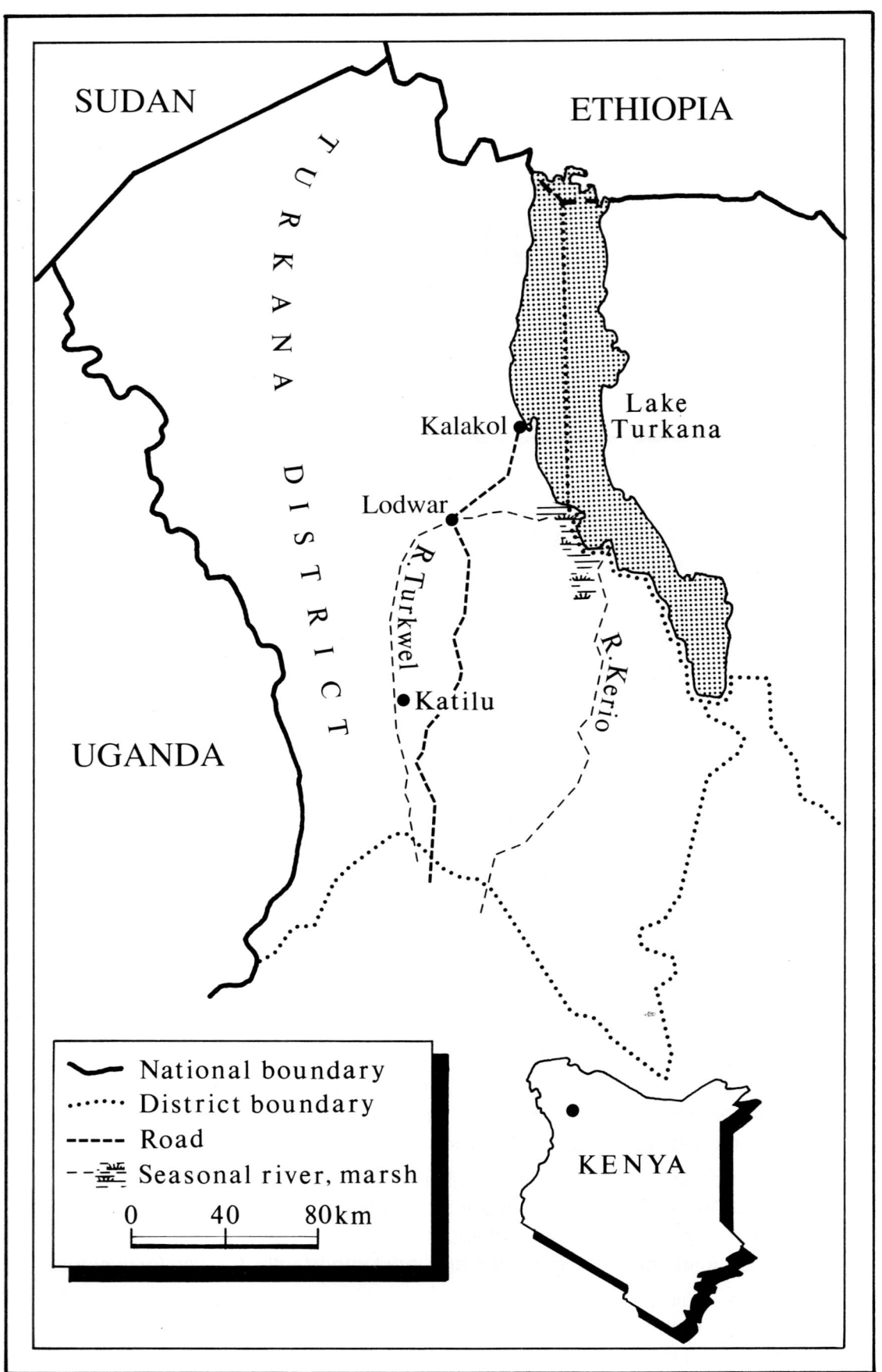

SUDAN

ETHIOPIA

TURKANA DISTRICT

Lake
Turkana

Kalakol

Lodwar

R.Turkwel

R.Kerio

UGANDA

Katilu

National boundary
District boundary
Road
Seasonal river, marsh

0   40   80km

KENYA

Name of project:	Turkana Rural Development Programme (TRDP)
Address:	TRDP, P.O.Box 175 Lodwar, Kenya. Tel: 0393 21026
Project area:	Turkana District
Average rainfall:	180-400 mm per year
Implementation:	Forest Department
Funding:	NORAD
	1981/84 - US$120 000 per year
	1984/87 - US$170 000 per year
	1987/90 - US$270 000 per year
Exchange Rate:	18 Ksh = 1 US$ (early 1989)

important tree species in the ekwar is *Hyphaene coriacae*. Its leaves are used to make the famous "Turkana baskets" which are exported to Nairobi and abroad.

Emerging problems

While land-use pressures in Turkana are not as severe as in other arid and semi-arid parts of Kenya, major problems are expected to emerge in the near future. A new tarmac road connecting the area with Nairobi has led to greatly increased cutting of the Acacia trees for charcoal in some of the riverine forests. Although the present rate of exploitation is relatively low the trend is worrying.

Other problems stem from the relief camps established to help drought victims. These have now turned into permanent settlements and there is great pressure on the natural resources around them. This has led to fuelwood shortages, lack of regeneration by the vegetation and a general over-stressing of the environment. The first signs of desertification are clearly evident near the settlements.

E. Barrow

Areas of riverine forest are also being degraded. Wherever the population pressure is high, little regeneration of the *Acacia tortilis* is taking place. The forests therefore consist increasingly of old trees and when these die or are cut there is nothing to replace them.

In the rangelands, away from the riverine forests, the *Dobera glabra* and the *Acacia tortilis* are also being over-exploited. *Dobera glabra* provides both fruit, which are cooked and eaten

The natural forest along the Turkwell River is more than a kilometre wide in places; use of trees is governed by traditional ownership rules.

163

by the Turkana, and browse during the dry season. Where the population density is relatively high, almost all the fruits are eaten so there are few seeds available for regeneration. Those that do germinate are killed by the goats. *Acacia tortillis* is also threatened by over-grazing, and in some places is being overcut for fencing. The Turkana prefer to use the whole young tree rather than selectively lopping the branches; in places this has led to eradication of the species.

Previous programme efforts

Concern over the plight of the Turkana dates back to the 1960s. Faced with serious drought and livestock losses, FAO embarked on a programme to develop irrigated agriculture along Turkana's largest river. The aim was to provide an alternative livelihood for pastoralists who had lost their herds.

E. Barrow

Canals, dams, pumps, airstrips and other structures were built and a range of new crops were introduced. After assistance was withdrawn in the 1970s the scheme more or less collapsed. The introduced crops are, however, still being cultivated in a few areas; in particular, a number of date palm varieties show promise and could provide farmers with a cash income.

Large numbers of microcatchments have been dug, enabling trees to be established in areas with less than 200mm annual rainfall.

There was also an attempt to promote tree growing. This was done on a food-for-work basis. About 500,000 microcatchments were dug all over the district and trees were planted. The results were far below expectations although a reasonable number of trees have survived.

One of the reasons for the comparative failure of the programme was lack of experience of arid lands afforestation among the project staff. Moreover, there was little consultation with local people about whether they wanted trees and which species they preferred. In many cases, the elders were not involved in decision-making. Some trees were planted in livestock routes and did not have the slightest chance of survival. The fact that people obtained payment in food meant that people were willing to carry out the work, but they had no real interest in the trees.

The present programme

The forestry component of the Turkana Rural Development Programme (TRDP) is being carried out by the Forest Department in Turkana District. Financial support and advice are being provided by NORAD. The first phase started in 1981 and the programme is currently in its third phase with funding up to 1990.

Tree planting was the main programme element in Phase I. Much of the work centred around infrastructure development, including the establishment of tree nurseries. An incentive scheme was set up under which families were given charge of planted areas and were obliged to water and protect the trees. They were paid KSh 1/= (about 5 US cents) per surviving tree per month. Typically, a family would manage 300 trees, planted in a dry, denuded area, and receive KSh 300/= (US$15) monthly until the trees were well established.

The people, however, turned out to be primarily interested in the money and were prepared to cheat the project where necessary. Thus when a tree died the family would replace it with one from a nursery so that they could continue receiving the money. Some staff also felt that the scheme discouraged spontaneous tree planting because it was felt that "TRDP pays you if you maintain one of its trees."

The net result of these activities in Phase I of the programme was a total planted area of only 20 hectares. It also became clear that arid land tree planting is a cumbersome, time-consuming and costly activity with a limited role to play in a vast region like Turkana.

From the beginning of Phase II the focus was, therefore, placed on working with the local people to identify their problems and trying to find ways of solving them. The project activities were broadened to include management of the riverine forests and rangelands which were being degraded.

Research was concentrated on the development of technical expertise specifically suited to the area and on locally relevant issues such as species which provide browse. The crucial necessity of creating a strong extension system, able to understand and actively involve the local people, was accepted. Phase III is continuing on basically the same lines, although with more emphasis on training.

Involvement of the Turkana

Bringing the local people into the discussions was an eye-opening experience for the project foresters. They had previously believed that the Turkana were solely interested in livestock and not in trees. They also knew nothing about the ekwar system and had assumed that the riverine forests were communally owned. The Turkana were therefore never consulted when the Forest Department imposed regulations on the use of the riverine forest.

The foresters also discovered that the Turkana, while seemingly uninterested in planting trees, have many traditional rules affecting the use of trees and shrubs. It became increasingly obvious that the Turkana do have strong interests in the forests, that they have a lot of knowledge about the trees, and that traditional management systems exist.

E. Barrow

The project recognises the need to establish a dialogue with local people; seminars are used to discuss ways of encouraging natural regeneration.

The Forest Department therefore decided to organise seminars at district and divisional level, involving district officials, extension agents, chiefs and other local leaders. During these, much more was learned about traditional rules and problems of over-exploitation in the various parts of the district.

As a follow-up, seminars are being organised at lower levels to include school teachers, important elders and leading women. By early 1988, some 70 seminars had been held and by 1990, the number will have reached perhaps 200. The objective of these is not only to raise the consciousness of local people, but also to learn about specific local problems and how to adapt rules and regulations to deal with them. It is also intended that a further series of seminars will be held to reach the next level of local people and officials.

The Forest Department staff have now built up a considerable body of knowledge and experience in the district. It is remarkable that the system of ekwars was not previously noticed during an anthropological research project in the district. It shows that foresters cannot automatically assume that social aspects of trees will be dealt with by other disciplines.

A typical local one-day seminar

Early in the morning, the forester, the chief and his assistants, and a group of some 40 other participants come together under a big *Acacia tortilis* tree. Clan elders, various teachers, some respected women, and staff of the Ministry of Agriculture are among them. After the introduction, the participants have a long walk through the forest and problems are discussed as the group comes across them. Typical items raised might include:

"Why are there no young trees in this forest?"

"What will happen if *Acacia tortilis* continues to be cut for charcoal?"

"Can the owner of this ekwar take any action to prevent further destruction of his plot?"

"How can we promote the valuable species?"

The seminar deals with issues which are specific to the location and tries to work out ways to tackle them. The meeting helps the forester to learn the local realities while Turkana participants are encouraged to define their own problems. Suggestions for possible action are made to the local community but they are aware that their ideas and opinions are also needed. In the end, it is up to the local people to make the decisions, not the forester.

A goat is slaughtered for lunch in the afternoon and this, together with the heat of the desert, makes everyone sleepy. After agreeing on future meetings, the seminar is dissolved.

Seminars lead to action to protect the local environment

The district and divisional seminars have become the forum for discussions on new legislation which integrates traditional Turkana rules and modern law. Once agreement has been reached on what should be done, the district authorities formalise the necessary regulations in new legislation. It is expected that laws which have been developed in this manner — by the people of the district, for the district — will be much more effective than if they had simply been imposed from above.

It has, for example, been agreed between chiefs, local officials and the Forest Department that charcoal export from the district should be banned. The ban has been in operation for some years now and is implemented by police and Forest Department staff together. The level of charcoal export has been reduced to almost nil. The fact that there is only one tarmac road leading out of Turkana has helped the implementation of the rule but even more crucial is that it is accepted by the local people.

In some areas, as a result of the seminars, the chiefs have become involved in other measures to protect the natural vegetation. In Lorugum, for instance, a large tract of rangeland was denuded of *Acacia tortilis* during the 1960s and 1970s. The chief and the elders have now imposed regulations to speed up the recovery. Under these, *Acacia tortilis* may still be cut, but only the side branches. This means that the dominant shoot is allowed to continue growing until it gets out of reach of the goats thus allowing the tree to survive. Currently, a spectacular regeneration of tens of thousands of young *Acacia tortilis* trees can be found in the area.

A project nursery; more seedlings are now going to schools and interested individuals.

167

Attempts to promote the regeneration of *Acacia tortilis* in the riverine forest and of *Dobera glabra* trees in the range have, however, not yet been successful. There are various suggestions about how this might be done and some isolated experiments have shown good results. For example, fruits of the *Dobera glabra* have simply been thrown into some thorny *Acacia nubica* shrubs where they are protected from grazing animals. The seeds have germinated and are growing well.

Tree planting becomes popular

Despite the relatively poor tree planting results of Phase I, it was decided not to discard this element completely. If nothing else, it is important to be able to show visitors that trees are growing and to keep funding agencies happy.

There has, however, been a major effort to broaden the scope of this element in the project. Rather than being planted by the Forest Department under the incentive scheme, an increasing proportion of the seedlings currently produced by TRDP go to schools and interested local people.

Planted trees can now be found in many family compounds. School compounds are also often well planted with trees as a result of a school competition organised by the Forest Department. The Department has also prepared a manual which will be sent to all teachers in the district.

In order to meet the increased demand for seedlings, production has increased greatly. The total number produced in 1981 was 25,000. This has now been raised to 900,000 per year coming from a total of some 20 nurseries.

Perhaps even more important than the numbers of trees being planted, is the fact that there has been a very significant change in people's attitudes as a result of the project's extension efforts. It is now clear that tree growing is becoming increasingly popular among the settled Turkana, a major change for people who never planted trees until recently. In one survey it was revealed that people were more aware of the Forest Department than any other extension agency; over 80% of those interviewed knew about it and some of its activities.

Locally focused research

The seedling survival rates in Phase I of the programme were poor because of inappropriate species and cultivation techniques. A considerable amount of research work has now been carried out to identify suitable species and their best methods of cultivation; this research will continue in Phase III.

The lack of rain in the area puts a premium on methods of reducing the amount of watering required. The use of microcatchments has been elaborately tested by the project and it has been found to be generally the best way to establish trees. The project's researchers have now drawn up a list of "proven species" for the area. A nursery manual containing guidelines for the watering regime and dealing with other technical issues has also been produced.

Responsibility for the research has now been transferred to the Kenya Forestry Research Institute (KEFRI). Research is also being carried out on the date palms introduced under the FAO irrigation project. The aim is to develop

168

methods by which these might be cultivated by individual farmers rather than in irrigated plantations. The results are promising and there is a good prospect that date palms will eventually be providing a cash income to some of the farmers of the region.

Spreading the message among foresters

The project foresters have shown a high degree of flexibility in their approach to the project. This has been a major element in its success. It is important that the message is spread among other foresters, particularly those undergoing their professional training.

Forestry education in the country is, however, still mainly focused on commercial plantation forestry in the Highlands. Until recently, there was no curriculum on arid and semi-arid lands forestry, nor were foresters trained in extension work.

The project has therefore organised courses for second-year forestry students of Moi University who now spend some time in Turkana studying arid lands forestry and the community approach to forestry extension. This is one of the most encouraging aspects of the project; it means that the hard-won experience of Turkana is being passed on to the next generation of Kenyan foresters.

Success story with uncertain future

It is obvious that the impact of the project on the whole Turkana District is relatively limited. Its effect is concentrated mainly in the areas where there is a certain amount of settlement and concentration of population. It is, however, in these areas that the major problems are found.

The project also demonstrates how far foresters may have to move from their original professional training in order to implement successful community level programmes. Virtually all the theories learned about species selection, spacing, forest management and harvesting had to be thrown overboard when working in Turkana.

The project thus reflects well on those implementing and funding it. The willingness to re-evaluate the programme and change direction after the relatively poor results of the first phase was crucially important. There are, however, severe doubts about whether it can continue unaided once the present funding is withdrawn.

At present, the Forest Department in Turkana is well-equipped with vehicles and has more than average operational funds, supplied by NORAD. Extension workers also have the use of four-wheel drive vehicles which is very unusual for extension workers elsewhere in the country, but almost a necessity in this vast district.

The Kenya Forest Department, as in many other African countries, has a low allocation of operational funds and would not be able to support the project at its present level. The fear is that, if NORAD withdraws, activities will virtually grind to a halt — unless government funds are greatly increased.

References:

BARROW, E.G.C. (1987a). "Extension and learning. Examples from the Pokot and Turkana, pastoralists in Kenya". Paper presented in IDS Workshop 'Farmers and Agricultural Research: Complementary Methods'.

BARROW, E.G.C. (1987b). "Establishment of trees in arid and semi- arid lands: some sociological and technical implications". Paper written for ODI Social Forestry Network.

TRDP (1987). "Forestry manual for primary school teachers in Turkana District". Draft.

TRDP (1988). "Project status summary report Phase I, II and III". Forestry Department.

EAST POKOT AGRICULTURAL PROJECT, Kenya

FARMING SEEN ONLY AS EMERGENCY OPTION BY PASTORALISTS

Livestock herding is the economic and cultural basis of the way of life of the Pokot pastoralists of central Kenya. This project found they were prepared to turn to farming as a survival strategy, particularly as it was supported by a food-for-work scheme, during a period of prolonged drought. But they were not interested in making such a change on a permanent basis and returned to pastoralism when the drought ended.

From famine relief to long-term development

East Pokot is situated in central Kenya in that vast stretch of arid and semi-arid land which reaches far into Ethiopia, Somalia and Sudan. It is part of the Great Rift Valley and the landscape is dominated by large extinct volcanoes and geological faults.

The annual rainfall is highly erratic. The year 1988, for example, brought 1300mm while other years bring disastrous drought. The annual average is 600mm and the potential evapo-transpiration is four times as much. The natural vegetation consists largely of dry Acacia scrub, such as *Acacia reficiens* and *Acacia mellifera*, interspersed here and there with riverine woodland. Drinking water is hardly, or not at all, available in some parts.

The area has long been inhabited by the pastoralist Pokot people who keep cattle, sheep, goats and camels. The land is communally owned but is broken up into traditional "group ranches" which are owned and managed by defined groups of people. The Pokot have a highly developed traditional knowledge of range management and the role of trees. Historically, there has been no need for tree planting since natural regeneration and range management have ensured that there are sufficient.

The security problems as well as the unreliable rainfall have frequently brought

171

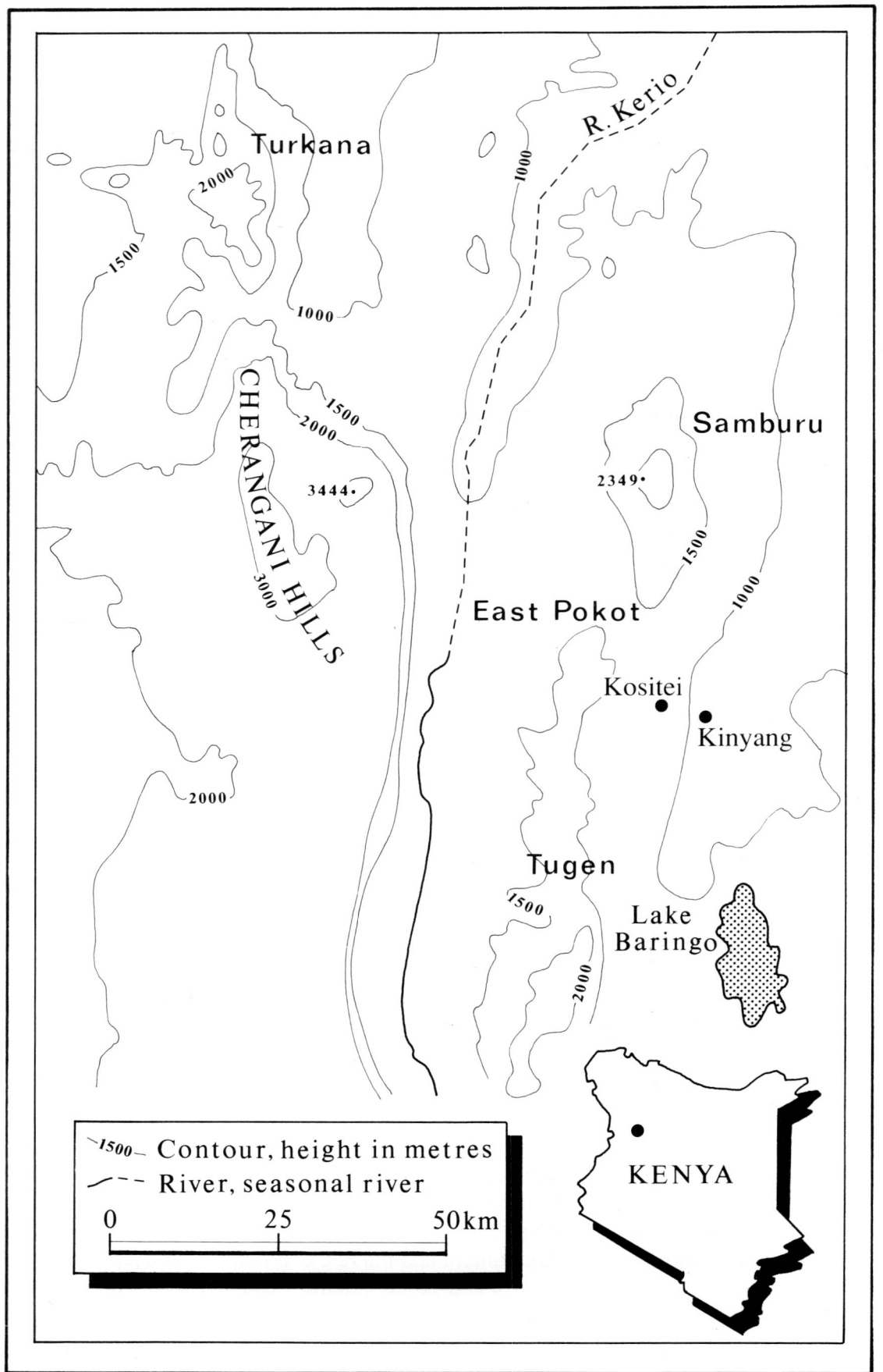

Contour, height in metres
River, seasonal river

0   25   50km

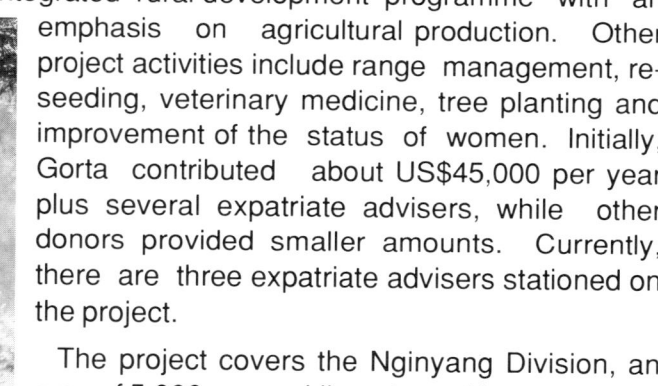

Name of project:	East Pokot Agricultural Project (EPAP)
Address:	Kositei Mission, Nginyang, P.O. Marigat, Kenya.
Project area:	Nginyang Division, East Pokot
Average rainfall:	600mm per year
Implementation:	Catholic Mission and Gorta
Funding:	Has varied between years, with maximum of US$45,000 in 1981
Exchange rate:	18 Ksh = 1 $US (early 1989)

the Pokot to the brink of devastation. One such occasion was in the 1970s when bad rains over several years caused famine and serious livestock losses. Several agencies, including an Irish mission, came to the area to provide emergency relief.

From 1976, the mission ran a food-for-work programme in which up to 1,400 people were employed. About 2,000 hectares were brought under cultivation using selective thinning of the natural woodland rather than total clearing. The food-for-work element was scaled down as the project gradually changed its emphasis from emergency relief to promoting long-term development.

This led to the establishment of the East Pokot Agricultural Project (EPAP) in 1978. From 1980, Gorta (The Irish Freedom from Hunger Council) has helped it evolve into an integrated rural development programme with an emphasis on agricultural production. Other project activities include range management, re-seeding, veterinary medicine, tree planting and improvement of the status of women. Initially, Gorta contributed about US$45,000 per year plus several expatriate advisers, while other donors provided smaller amounts. Currently, there are three expatriate advisers stationed on the project.

The project covers the Nginyang Division, an area of 5,000 square kilometres with a population of about 30,000. The population density averages 6 people per square kilometre but varies considerably

From the beginning, a basic premise of the project was the need to develop agricultural production since, under the conditions of the late 1970s and early 1980s, it was clear the Pokot could not support themselves on pastoralism. Tree planting was also seen as an important element in the restoration of the degraded rangelands in Pokot. This was reinforced by the national support for tree growing given by the President of Kenya.

Paul Kerkhof/Panos Pictures

The project nursery at Kositei; demand for seedlings has been disappointing and project staff are thinking of phasing out the nursery.

173

Development of a package

When the project began, there was little local knowledge of tree planting on which it could build. Neither was there much information at a national level since most tree growing research has been concentrated on the high and medium potential areas rather than the arid and semi-arid lands.

A number of research plots were therefore established to screen species for their suitability for the harsh climatic conditions and to arrive at the most appropriate cultivation methods. The best results were achieved with a model derived from the Negev in Israel and using water harvesting by means of semi-circular microcatchments.

The semi-circle is dug with a hoe and has arms three metres in length. The tree is then planted at the lowest point where the run-off rainwater has collected and infiltrated into the soil. One of the most attractive features of the microcatchments is that they are simple and any farmer can make them.

Some 50 tree species were used in the trials, including a range of exotics. Among those which showed the best growth were *Azadirachta indica*, *Acacia tortilis*, *Atriplex nummularia*, *Balanites aegyptiaca* and various *Prosopis* species.

Data collected on local attitudes to trees indicated that they were regarded as highly important by the Pokot. There was, however, no survey of how people felt about the totally new concept of planting trees.

Paul Kerkhoff/Panos Pictures

Livestock are excluded from the area in the background, which is protected by a thorn fence and paid guards; where controls have worked, natural regeneration has been encouraging.

Training and extension

When the project began, extension was seen as a two-way process with the project learning from the local people, especially over the question of range management. EPAP developed a small extension team, and a total of about 10 local Pokot were selected and sent from time to time to an agricultural training college.

Although agricultural development was the main thrust of the project, tree planting was also considered important. Seedlings were grown in the project nursery with paid labour and under the supervision of EPAP staff. Annual seedling production began at several thousand and grew to several tens of thousands but is now declining.

Most seedlings have been used for planting in project lands and public places, though some have been locally sold at a subsidised price. Where individual Pokot have bought seedlings, they have been mainly fruit trees. The majority of the seedlings now produced are not used locally but are dispatched to the highlands outside the project area.

EPAP established five tree plots to demonstrate tree growing techniques. These had areas up to several hectares and were fenced with Acacia thorn branches and protected by guards under the food-for-work scheme. They have worked well as trial plots and have also provided large quantities of tree seed much of which has been sold throughout Kenya and provided a significant revenue for the project. The plots have had little, if any, effect in persuading the local people to establish plots of their own but have, perhaps, helped raise local awareness of the potential for environmental management.

Perhaps the greatest interest in tree planting has been in the few primary schools in the area. Individual households were also encouraged to plant trees, but after more than a decade of promotion, there is still little interest. "I would be hard pushed to show you one family compound with successful tree planting," said one of the project staff. Agricultural demonstration plots were also established but have all been abandoned because of lack of local interest. As one project worker put it: "The final test was handing over a well-established demonstration plot, with water harvesting systems and everything. None of the local people were interested, nobody took it."

In some areas, where the availability of water is relatively good, there has been a somewhat better response but it is rarely enthusiastic. A member of the project staff said: "It has been a struggle all the time to sell tillage and tree planting to the people. But when I talk to them about goats and camels they are with me."

The government extension services in the area have recently been expanded and the project has cut back on its own extension activities. EPAP had hoped the government would absorb its extension agents, but this turned out to be wishful thinking. This is a common experience among NGOs in the country. It is virtually impossible to hand over to the government any extension service which has been built up with project funding.

Experiments in range management and controlled grazing

The major problem for the Pokot is, in fact, the low productivity of their pastoral system. This is mainly a result of over-grazing but has been aggravated by security problems which have forced the herders to retreat into confined areas, away from the borders with other tribes. Some border areas have long been "no man's land". The Pokot have traditionally controlled grazing in accordance with a set of rules interpreted and administered by the elders. In recent years, the authority of the elders has been gradually eroded and there has been less compliance with the grazing rules. In order to emphasise the importance of the reserved grazing areas and bolster the authority of the elders, the project provided labour and materials to control grazing on certain hills.

The largest controlled area covered 3,000 hectares. It was fenced off with thorn branches and guarded by paid workers. It was reseeded with grass using *Cenchrus ciliaris, Enterpogon macrostachys, Eragrostis superba* and a number of other species. The effect on range quality was dramatic and the impact on wood production was greater than any likely tree growing programme.

At certain times, as determined by the elders, limited numbers of livestock were allowed into the protected areas for a small fee. The system worked well for some years and was praised in the national press. But when the drought in 1983, the controls collapsed and have not yet been restored.

Difficulty in promoting sustainable innovation

The East Pokot pastoralists have traditionally had a livestock economy and culture. This makes considerable sense given the erratic rainfall pattern of the area; in bad years the herds can be taken to where fodder is more likely to be available. It has also meant that livestock ownership and production are highly valued whereas crop cultivation is traditionally somewhat despised.

During the droughts of the 1970s, however, the Pokot could not subsist on the production of their animals and theirproblems were compounded by their hostile neighbours. Rather than simply providing emergency relief, the project staff, understandably, were looking for a more sustainable solution and crop production seemed preferable to pastoralism. Tree planting was included in the package because it was widely seen as the right way to stop desertification and avoid the "fuelwood trap".

Subsistence crop production under the climatic conditions of East Pokot is, however, always likely to be difficult and risky. When the project wanted to hand over its responsibilities to the Pokot, it found they had little interest in becoming agriculturalists. With the return of the rains and the recovery of the livestock herds in the 1980s, they simply wanted to resume their pastoralism.

The controlled grazing component of the project seemed, however, to fit with the pastoral tradition and temporarily had a significant impact. The fact that it collapsed under the intense local pressure which built up during the drought period is not, in itself, surprising; pastoralist societies such as the Pokot often set aside certain areas for emergency grazing during droughts.

The reason the grazing controls were not restored after the drought is a matter of some debate. It may simply be that the authority of the elders was not sufficient to ensure that the lands in question were withdrawn from use once the need for them had passed. Another possible reason, or contributing factor, may have been that people felt that, since the project had previously taken responsibility for the area and provided fences and guards, it should continue to do so.

Experience in the project shows that the Pokot are primarily interested in livestock herding. It is their traditional strategy for survival under the difficult conditions in which they live and they are open to measures such as veterinary medicine, water supply and range management which improve food security and help them increase production. Agriculture is an entirely different matter. The project's food-for-work schemes provided them with a means of survival when there was no alternative available. But this remains a considerable way short of any long-term commitment to agroforestry or the agricultural way of life.

References:

EPAP. "Annual reports of the East Pokot Agricultural Project, 1978-1980, 1981, 1983".

176

FOREST LAND USE PROJECT, GUESSELBODI
Niger

COMMUNITY PARTICIPATION WORKS BUT THE COSTS ARE HIGH

Large areas of the Guesselbodi natural forest outside Niamey have been destroyed by fuelwood cutting and over-grazing. Attempts to combat the destruction with eucalyptus plantations and other conventional forestry measures proved costly and ineffectual and have been abandoned.

Much greater success has been obtained by involving the local community in managing the natural forest for the production of wood and fodder. But at present fuelwood prices, even this approach is far from covering its financial costs. It appears as if such programmes cannot be sustained without substantial and permanent donations from abroad.

The vanishing forest

The Guesselbodi forest is in the south west corner of Niger only 30 kilometres from the capital city Niamey. The area is part of the high plateau which covers most of the country. The climate is Sudano-Sahelian with a rainfall of about 500mm per year which falls between May and September.

In many classification systems the Guesselbodi vegetation would hardly be labelled a forest. There are scattered trees of *Combretum micranthum*, *Combretum nigricans*, *Boscia senegalensis*, and some acacia shrubs. There are also some *Prosopis africana* and *Commiphora africana*. *Cenchrus bifloris*, *Aristida mutabilis* and *Eragrostis tremula* are the major grass species.

The region is mainly inhabited by Djerma farmers but the semi-nomadic Fulani and Tuareg herders pass through from time to time. The average population density is about 20 people per square kilometre but most people are settled along the rivers. In the higher altitude dry areas, the population density can be as low as 2 people per square kilometre. The Guesselbodi classified forest itself is not inhabited.

177

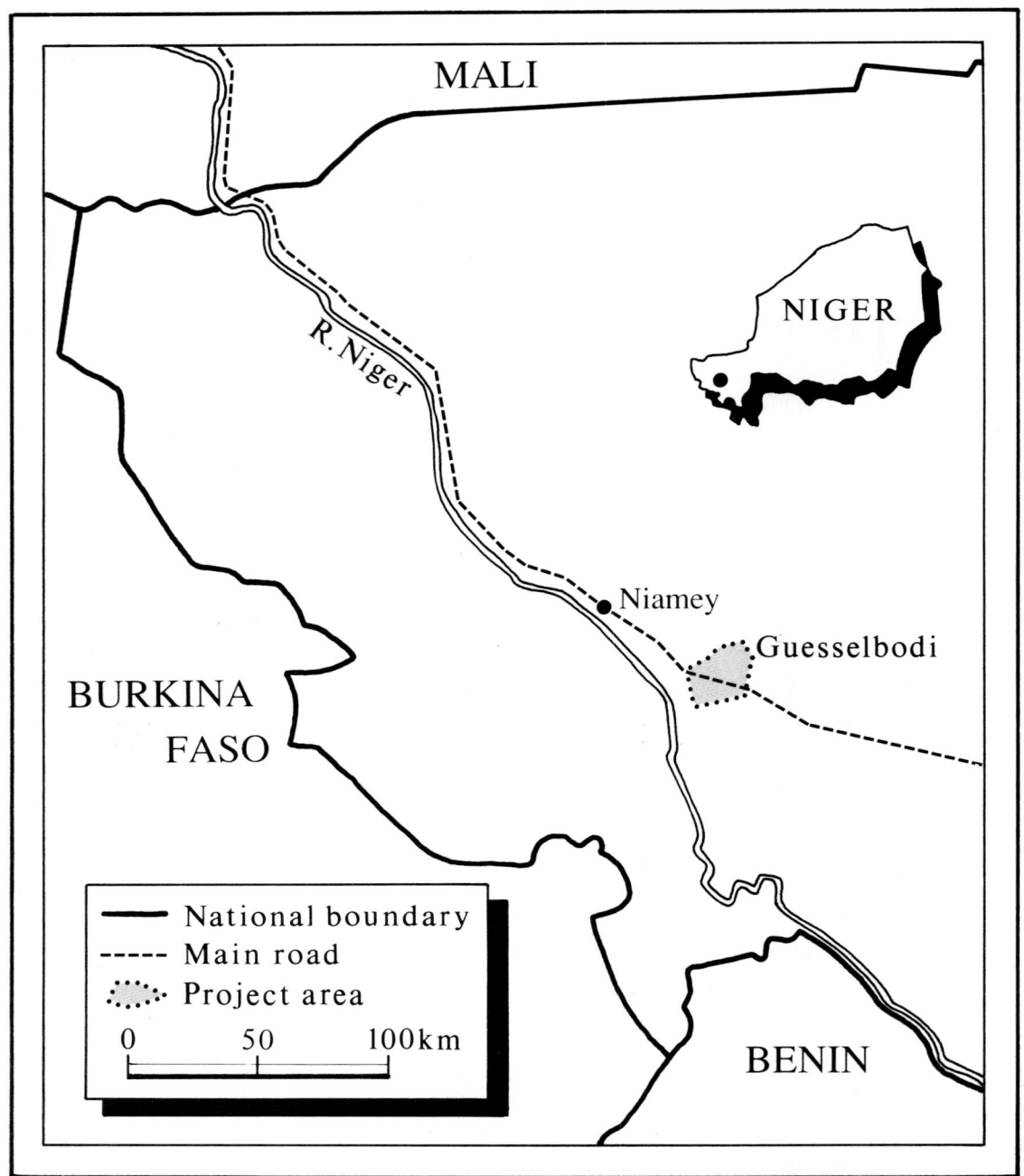

Name of project:	**Forest Land Use Project/Guesselbodi (FLUP)**
Address:	**B.P.12203, Niamey, Niger.**
Project area:	**Guesselbodi Forest**
Average rainfall:	**500mm per year**
Implementation:	**Forestry Service with expatriate advisors**
Funding:1980-1988:	**US$300-600,000 per year from USAID**
Exchange rate:	**300 FCFA = 1 US$ (early 1989)**

The city of Niamey has about 700,000 inhabitants and is growing at the extremely rapid rate of 10% per year. Firewood is the only source of fuel for over 90% of the city's households. Niamey is among the poorest cities in Africa and is likely to remain dependent on firewood for a long time.

The Guesselbodi forest was gazetted in 1948, and aerial photographs have been taken regularly since then. These show that the tree and bush cover have decreased at a rate of 2-3% per year, or by about 50% over 30 years. Much of the land is now entirely bare and covered by a laterite crust which allows no natural regeneration. Even within the best parcels, vegetation is usually sparse and in sloping areas there is often severe gully erosion.

The FLUP project

From 1971 onwards, organisations such as the World Bank, CTFT (Centre Technique Forestier Tropicale) and CCCE (Caisse Centrale de Cooperation Economique) of France, funded attempts to establish forestry plantations in Guesselbodi. The Forestry Service planted several hundred hectares mainly with eucalyptus but ran into the same problems encountered in conventional forestry programmes elsewhere in the Sahel. The establishment and maintenance costs were extremely high, seedling survival was poor, and the growth rates were far lower than predicted. As a result, this approach was abandoned.

In 1980, the Forest Land Use Project (FLUP) was established; in French it is the Projet Planification et Utilisation des Sols et Forêts (PUSF). The objective was to promote the regeneration and management of the forest on a locally sustainable basis. Multiple use of forest lands was to be encouraged, with the best soils being reserved for crops rather than trees, and other areas used for grass production. A key aim was to involve the local people as much as possible.

The project studied all the major forest lands inside a radius of 100 kilometres from Niamey and identified eight areas for more detailed investigation. Guesselbodi was selected for a pilot project to develop and prove the proposed new management system. Although some work has been carried out in the other seven forest areas, Guesselbodi is by far the most advanced.

The project was attached to the Forestry Service in the Ministry of Hydrology and Environment and has been financed from the beginning by USAID. The budget has varied from about US$300,000 to US$600,000 per year for the whole project, including the other forests. Technical assistance has been provided by up to four expatriate advisers and several volunteers. The project is currently in an interim phase in which the funding is lower than in the past but negotiations are underway to renew it at a higher level.

Trials and planning

In the early stages of the project a management plan was worked out. The total area of 5,000 hectares was divided into 10 parcels of about 500 hectares to be exploited in consecutive years thus giving an overall rotation period of ten years.

179

It was found that only about 3,000 hectares could provide any worthwhile yield of fuelwood with the remaining 2,000 hectares being seen as beyond salvage for wood production. Even the productive area was in bad shape, with only 207 hectares having a crown cover of 80-100%.

Silvicultural trials and experiments were held to test the proposed management measures. These took several years and some are still under way. They showed that Combretum species were the most successful for

Acacia holosericea *planted in a V-shaped microcatchment. on what used to be barren land.*

coppicing and found that a considerable increase in wood production occurs if the trees are coppiced at a height of about 30 cm and protection is provided against browsing. In some trials where no protection was provided, the survival rates after coppicing went down to 10%.

Direct seeding of *Balanites aegyptiaca*, *Bauhinia* species, *Acacia* species and others was repeatedly tried but without success. Tree planting in microcatchments was technically successful. So also was grass reseeding using a variety of species. The project finally decided that apart from *Acacia holosericea*, an Australian species which grows very quickly under local conditions, only indigenous tree species would be used for planting.

Trials were also carried out to assess the effect on natural regeneration of mulching the soil with twigs and branches, and of tilling the soil by breaking up the surface layer with a pick axe. Mulching was particularly effective in stimulating regeneration, but the best results were obtained when the two methods were combined (see Figure 19.1).

The beneficial effect of mulching is thought to be related to the increased termite activity that mulching stimulates. The termites create fine tunnels which improve the water penetration in the soil. This phenomenon seems to be similar to the "Zay" method which has been developed in Yatenga, Burkina Faso (see PAF project profile). The presence of mulch on the soil surface also helps trap wind-blown seeds.

An extensive and costly forest rehabilitation package

Under the project, selected areas have been planted with trees and grasses. Protection against grazing has been found necessary for at least three years after tree planting. Guards are hired by the project at a rate of one per 250 hectares; usually they are Tuareg who have no ties with the local farming communities.

Each guard is equipped with a camel and a sword. Animals found in the protected areas are impounded and the owner has to pay a fine before they are

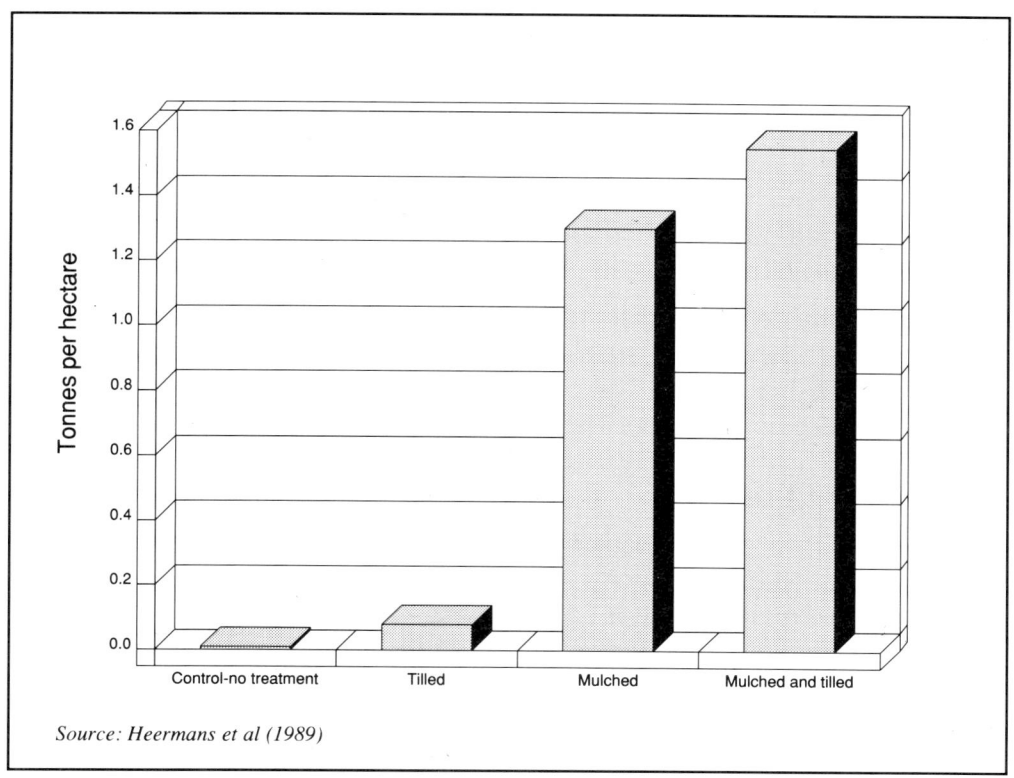

Source: Heermans et al (1989)

Figure 19.1 Effect of tilling and mulching on natural regeneration. Graph shows total biomas production after 3 years (see text for details).

released. This system of control is extremely effective and is cheaper than fencing. Nevertheless, it is still costly as each guard is paid FCFA40,000 per month (about US$130) and the costs of the camels and their maintenance also have to be borne. At present, there are 12 guards employed in the forest.

In rehabilitating areas where there is no longer any woody vegetation, contour bunds were constructed to reduce sheet erosion. Whenever the materials were available nearby these were made of rocks; otherwise earth bunds were constructed. The costs of treating an area in this way were calculated to be about FCFA68,000 per hectare (about US$230). In hilly areas, check-dams were built to reduce gully erosion.

Microcatchments were constructed by hand in areas scheduled for tree planting. They are in the shape of a "V" with arms 2.5 metres long. The apex of the "V" is slightly higher than the end of the arms so that water is drained away without overflowing and damaging the wall.

Microcatchments in sandy soils turned out to be ineffective; they fill up quickly with wind-blown sand and, in any case, there is little water to catch. In very heavy soils, on

Paul Kerkhof/Panos Pictures

Tuareg guards have proved highly effective in patrolling the Guesselbodi forest.

the other hand, the water cannot drain away and it may be necessary to choose tree species which are resistant to water-logging.

Where a laterite crust had developed, bulldozers were used to break up the soil to a depth of 40cm. The costs were extremely high, but the results were generally poor and the technique was dropped. Breaking the laterite crust with a pickaxe turned out to be cheaper and the grass cover returned in the first season, but the soil reverted to its original crusted state after two more seasons.

Every year the central nursery of Guesselbodi has produced about 100,000 seedlings which have been mainly used for planting in the microcatchments. Encouraged by trial results, the project decided to apply 30 grams of NPK fertiliser and 39 grams of Super Triple Phosphate to each planted tree. Weeding was done once in the planting season; repairs to microcatchments and bunds were made at the same time.

A large amount of grass reseeding has also been carried out with about 5,000 kilograms of seed of the species *Andropogon gayanus* being sown annually. Because of the many barren areas in Guesselbodi, bush fires do not pose a threat and firebreaks are unnecessary. But they are needed in some of the other forest areas under scrutiny by the project.

The result of these measures is that the annual sustainable fuelwood production per hectare has gone up by a factor of two to three, increasing from about 0.5 steres, or stacked cubic metres, before the project, to 1.0-1.5 steres. The grass production has also gone up and has reached an average of 640 kilograms of dry matter per hectare per year (Heermans et al, 1987).

Towards participatory forest management: a difficult step

A socio-economic survey, using a questionnaire prepared in the local language, was conducted to establish how the forest was used by the local people and how they would react to the planned interventions. This revealed that the surrounding population used the forest in a variety of ways including livestock grazing, honey collection, and as a source of medicines; it also found that there were certain traditional pathways which ran through the forest.

The project held a number of local meetings in an effort to involve the villagers in the management of the forest. The villagers proved uninterested and distrustful of the Forestry Service. This was because foresters have traditionally had a repressive role and seen their task as protecting the forests against the local people; in one of the local languages, foresters are referred to as "Sarkin Daji" which means king of the forests.

The problem of lack of effective communication between foresters and local people is well illustrated by some of the communal forestry projects carried out in the area. In some of these, although the trees have been planted, neither the Forestry Service nor the villagers claim them and they simply grow and die. The local expression is that "the trees belong to the earth".

Another major obstacles to local participation turned out to be the fact that the villagers are not traditionally involved in cutting fuelwood for the market. This has always been done by traders, "people from outside". But local interest in the project greatly increased when the villagers were offered tangible benefits in return for accepting responsibility for forest management.

The government of Niger has a national policy of encouraging the cooperative movement and there are cooperatives in many villages, though some may be dormant. It was thus logical for the project to set up a forest products cooperative embracing the nine villages bordering the Guesselbodi forest.

Work on setting up the cooperative began in 1985/86 and in February 1987 it signed a contract with the Forestry Service, through the Minister of Hydrology and Environment, to organise commercial fuelwood exploitation in the area. This was a big step for the Forestry Service since it delegated some important responsibilities in the management of gazetted forest to local people. A further clause covering the production of hay was added later.

Under this contract the roles of the Forestry Service, the individual villagers and the cooperative have been defined as follows:

- The Forestry Service is responsible for forest rehabilitation and conservation. It is ultimately responsible for the annual harvesting schedules and it sets the rules for harvesting. It carries out the necessary policing.

- The cooperative is accorded the right of forest exploitation subject to the rules laid down by the Forestry Service.

- The residents of the neighbouring nine villagers are members of the cooperative and have the right to buy permits to harvest fuelwood or hay. Members have to sell all wood to the cooperative, apart from the amount they need for home consumption but can sell hay as they wish.

- The cooperative sells the fuelwood it purchases from its members to licensed fuelwood merchants.

According to the Forestry Service rules, dead and dying trees can be harvested, and live trees of specified species and diameter can be cut. The woodcutter is given a small measuring tool, made of a stick with nails driven into it, to determine which trees can be cut and which are still too young. Trees which are particularly needed to produce seed for natural regeneration, or are important for wildlife, are marked by forest technicians and may not to be cut.

The proceeds from the fuelwood sales are shared, with 75% going to a Forestry Fund, managed by the Forestry Service, for the rehabilitation and maintenance of the forest, and 25% going to the cooperative. The fees for fuelwood cutting permits also go to the Forestry Fund while those for cutting hay are retained by the cooperative.

A fuelwood collection point, managed by the cooperative, has been set up on the main tarmac road to Niamey. There, merchants' vehicles buy the fuelwood and take it to the city. The cooperative only sells wood in the rainy season, when prices are at their best.

In 1987, the first year of fuelwood cutting under the project, 980 steres of wood and 4,355 bundles of hay were harvested. This produced a revenue of FCFA478,000 (US$1,600) for the Forestry Fund and FCFA128,000 (US$430) went into the Cooperative bank account. In 1988, a total of 1,822 steres and 44,804 bundles were harvested and the cooperative earned about FCFA2 million (US$6,700). In 1989, the number of registered woodcutters was 149 and a total production of 2,400 steres was expected.

In setting up the cooperative, and ensuring that it obtains tangible benefits from the project, FLUP has made an important step towards involving the local people. Nevertheless, it is still the Forestry Service which is in charge of the forest. A project survey revealed that the villagers see the function of FLUP as controlling the forest exploitation, with the cooperative organising the marketing of the

The firewood collection point on the road to Niamey; the wood is kept until the rainy season, when prices are highest.

fuelwood to ensure a good price. The villagers saw their own role as complying with the rules. Women play virtually no part in the commercial exploitation of the forest.

The problem of financial viability

The biggest problem facing the project is that it is far from being financially viable. Analysis shows that the cost of full rehabilitation, including micro-catchments and tree planting, can reach FCFA200,000 (US$670) per hectare. A study carried out with the assistance of GTZ showed that this cannot be covered by the revenue from fuelwood sales in Niamey at present prices (de Winter et al, 1988).

Raising fuelwood prices is hardly a practical option since the cooperative is only supplying a small proportion of the fuelwood used in Niamey and cannot raise its prices above those of other suppliers. At present there is no sign of any general increase in the price of fuelwood in real terms. In any case, the GTZ study suggests that fuelwood prices would have to be 3-4 times higher than at present if the project were to show a positive financial return. At that level, it is doubtful if fuelwood would still be competitive with other fuels.

Neither is it possible to envisage any dramatic increase in the output from the forest beyond the extra 1-1.5 steres per hectare per year already achieved. Hay production could probably be raised, but the financial effect would be minor. It has been suggested that the fuelwood might be sold directly to consumers in Niamey, thus cutting out the merchants' profit, but it is unlikely the cooperative would be more effective in production and distribution than private enterprise.

There have also been some moves to subsidise the project through the existing tax on fuelwood. Traditionally, this has been levied on all fuelwood loads entering Niamey. A contract was recently signed between the Government of Niger and the World Bank under which the tax on fuelwood entering Niamey will be substantially increased. The idea is that the revenue raised in this manner will be used to subsidise the project.

There are, however, doubts in practice and principle about any such arrangement. Many fuelwood consumers are extremely poor and there is likely to be considerable resistance to any major increase in fuelwood taxes. Moreover, the Government will find it difficult to continue using such taxes to subsidise the project when there are so many other urgent calls on its very limited resources.

An alternative approach is to cut the costs of the project. It is understandable that the Forestry Service likes to to include soil conservation and tree planting in its programme of forest rehabilitation and management. But analysis shows that the returns from such activities are low in the harsh climatic conditions found in the project area.

The project has shown what can be done by simply providing the Tuareg guards. Their presence has stopped unchecked wood cutting and grazing and it is expected that this measure by itself will lead to considerable regeneration of the forest. The GTZ study suggests, however, that such protection should only be provided in those areas which have a reasonable potential for spontaneous regeneration.

What is abundantly clear is that it makes considerably more sense to conserve and manage forest resources while they still exist rather than trying to restore them with costly soil conservation and tree planting measures after they have been destroyed or severely damaged.

Where next?

FLUP has successfully and practically demonstrated that natural forest and bushland in a semi-arid environment can be rehabilitated. In Guesselbodi, wood and grass production have been restored, wildlife has returned and in many other ways the forest has started functioning again. Moreover, this has been done in a way that benefits the local villagers.

The project, however, continues to face severe practical and financial obstacles. Accounting problems in the cooperative have been reported and it is still under the guidance of the Forestry Department. Indeed, it remains uncertain whether the cooperative is the best way to organise local participation in forest management.

It is also unclear whether forest management programmes, however they try to cut their costs and involve the local community, can ever hope to be financially viable in the harsh conditions of the arid and semi-arid regions where they

Paul Kerkhof/Panos Pictures

Transporting hay to Niamey; firewood must be sold to the cooperative, but villagers can collect and sell hay themselves.

are required. For the present, it is certain they can only survive in a poor country like Niger with the help of substantial external funding.

References:

DE WINTER, J. et al (1988). "Etude Aménagement et protection des forêts naturelles dans la région de Niamey." Deutsche Forstinventur-Service GmbH, Federal Republic of Germany.

FLUP. "Annual Reports, 1982-86".

HEERMANS, J., G. Minnick and C. Polansky (1987). "Guide to forest restoration and management in the Sahel based on case studies at the national forests of Guesselbodi and Gorou-Bassounga, Niger". FLUP/USAID.

PART III

APPROACHES TO PROJECT DESIGN AND IMPLEMENTATION

APPROACHES TO PROJECT DESIGN AND IMPLEMENTATION

Many different approaches have been used in the design and implementation of agroforestry projects. Most projects have had to learn through experience which are best suited to the area in which they are working, and which are most realistic, given the resources available to them.

Drawing upon this experience, some of the key aspects of the design and implementation of projects are highlighted in this section. Different approaches are compared and some of the main dilemmas facing projects are discussed.

INITIAL SURVEYS

Carrying out an initial survey to find out about peoples' attitudes and priorities, and to develop an understanding of the local community, is an obvious first step. Such a survey also provides a baseline against which the project impact can later be measured. In practice, however, many of the projects visited failed to carry out a survey, or only did so a number of years after the project began.

It is easy to understand why surveys are not done. They take time and money and there is often a great deal of pressure on projects to show they are having an immediate impact. Project staff may also feel a survey is unnecessary because, they say, "we know this area". But in almost all cases where surveys have been carried out they have yielded valuable and unexpected results. And in the long term they have saved a lot of effort and resources.

Among the projects covered, there have been two main types of surveys: the general socio-economic baseline survey, and the diagnostic survey. The first is broader and seeks a range of information about the people and activities within the project area. It includes population data and information about agriculture, livestock and forestry. Diagnostic surveys are normally more focused, and concentrate on specific constraints in the farming system which are related to trees. This approach has been developed most extensively by ICRAF in its Diagnosis and Design (D&D) methodology.

The majority of agroforestry surveys are based on interviews with local people. A few projects have used more elaborate techniques such as aerial photography and detailed vegetation mapping. These have yielded interesting results but the high costs involved put them out of reach of most projects.

The commonest survey method is to use a formal questionnaire consisting of a standard list of questions prepared in advance. The advantage of asking everyone the same questions is that processing the information is relatively easy, and it allows quantitative conclusions to be reached.

If an attempt is made to deal with too many issues, however, the list of questions can become unmanageable. The University of Dar es Salaam used a questionnaire of 48 pages to carry out a survey for the SECAP project in Tanzania. In such cases the patience and concentration of respondents can be severely strained. Data processing becomes complicated and requires access to computer facilities. The final reports also tend to be unwieldy because of the overload of information.

Surveys have shown the farmers often know far more about tree growing than projects had assumed; this Tanzanian farmer is collecting naturally germinated **Grevillea** *seedlings for transplanting to his backyard nursery.*

Some subjects are hard to investigate with a standard questionnaire, and require a more flexible and informal survey approach. An example is the cultural survey carried out by KWDP in Kenya, which assessed how husbands and wives discuss fuelwood issues in the family. Such surveys can yield extremely useful information. The disadvantages are that they need highly-qualified survey staff, and, since only a small number of people can usually be sampled, the results may not be properly representative.

Because they are difficult to address, questions such as land tenure, tree ownership rights, and who makes decisions about trees on the farm, are often omitted in surveys. But they are crucial to the design of projects. Time and again, it appears obvious from the outside that farmers should want to plant trees. But, in practice, factors such as uncertain land ownership, male migration, and the weak position of women on their husband's farms prevent people from doing so.

When surveys are carried out it is important to ensure that they are seen as an integral part of the project. In some cases surveys have been treated solely as a planning tool with project staff having little or no involvement. They have been carried out as a separate exercise, sometimes by an external consultant.

The final product has tended to be a lengthy academic report which has been of little use or interest to project staff.

In other cases, surveys have been more closely integrated into the project. Extension staff have been involved in planning the questionnaire and have themselves worked as enumerators. By visiting farms in different locations they have learned about the variations between them, and gained a better appreciation of local knowledge. In some projects they have also assisted in the data processing. All this has helped extension staff to feel involved with the survey and more interested in its results. While it may be slower and more cumbersome, this can have considerable educational value and bring long term benefits to the project. It also helps to make sure that the right approach is taken when dealing with sensitive issues.

A problem with many surveys is their size. The larger the survey, the higher the cost to the project, and the more difficult it is to involve extension staff. Some have covered

Geoff Barnard/Panos Pictures

Surveys carried out by KWDP in Kenya found that there are important cultural constraints limiting the rights of women to plant trees.

hundreds, or even thousands, of farmers and have become so elaborate that it has been difficult to draw clear and usable conclusions. In the extreme, the amount of work involved in the data processing has been so great that it did not leave enough time to write up the final results properly. On the whole, experience shows that surveys which concentrate on a few important problems and potential solutions, are generally more useful and cost effective.

Research: working towards solutions

Because of the lack of "off-the-shelf" agroforestry techniques, most projects have had to incorporate a research component. Among project staff there has been a good deal of apprehension about this. Their experience is that research programmes often produce inconclusive or contradictory results, and that they tend to raise far more questions than they answer. Designing a research programme so that it can produce usable results quickly and economically is one of the most difficult tasks in a project.

The research methods that have been used have varied widely. Some projects have carried out "on-station" research, in which trial plots have been established under controlled conditions, either at existing research centres, or at the project headquarters or main nursery sites. These have been used to screen different tree species; trials have also been carried out to assess the impact of trees on foodcrops and test various agroforestry combinations.

On-station trials are easier to monitor than on-farm trials and provide a useful focus for visitors; but there are often problems in transferring the results to the field.

With trials of this kind, care has to be taken that there are proper control plots and enough replications to ensure that statistically valid results are obtained. A number of projects have used recommendations developed by ICRAF in designing their experiments.

A feature of on-station trials is that they require considerable resources and the long term commitment of research staff. Some projects have embarked on trials involving dozens of agroforestry combinations without the necessary long term view. They have therefore produced very little in the way of statistically meaningful results. Even so, they may not be a complete waste; they can give some useful guidance on species behaviour, and provide a convenient place where guests can be shown around — an important consideration for projects that receive large numbers of visitors.

The major disadvantage of on-station research is the difficulty of translating the results into practical recommendations for farmers. Almost inevitably, the conditions found on a research station are different from those on a local farm. There is, therefore, a widespread feeling among project staff that more emphasis should be placed on on-farm research in which trials are carried out under practical farming conditions. Since this is a relatively new concept in agroforestry, and even in agricultural research, there are few guidelines to work by and most projects have had to develop their own methods.

KWDP in Kenya, for example, used on-farm research to investigate the performance of fuelwood species and to find out what farmers felt about their use in the very densely populated project area. The project used an open approach and the 22 farmers chosen were allowed an almost entirely free choice in how the trials were carried out. The project provided the seedlings and described their characteristics and cultivation needs, but the farmers themselves specified the number they wanted. The decision on where and how to plant the seedlings was also left to the farmers.

The survival and growth of the seedlings was monitored every six months. Farmers were also interviewed about their views on the management and use of the trees. How interested they were in the trial species could be deduced from where they planted them, how much attention they gave them, and the

Geoff Barnard/Panos Pictures

extent to which they were prepared to set up nurseries to produce their own seedlings. Another point about the trials was that they involved project staff. This meant that they identified with the study, rather than seeing it as something separate from their work.

The major problem with such open on-farm trials is that it is extremely difficult to obtain quantitative data because of the differences in the way farmers treat their trees. Some carry out frequent weeding, others none; some protect the seedlings rigorously from animals, others do not; trees may be planted as woodlots, hedges, on rocky ground or amongst crops. In 1988, the project staff concluded that this loosely structured approach would have to be supplemented with more closely controlled on-farm trials if quantitative data were to be obtained.

The PAFSAT project in Cameroon provides a good example of such controlled on-farm trials. Here, 80 standard trials are being carried out on selected farms to test the proposed project ideas and compare them with the traditional cropping system. Initially, project staff were heavily involved in monitoring and controlling the work done in the plots but this was found to be too time-consuming. The farmers have now been trained and given responsibility for the implementation and measurement, but not the design, of the trials.

In principle, controlled trials of this type should lead to statistically valid results and quantitative performance data. But even here, the results from about half the 80 farms have had to be discarded because of uncontrollable variables, such as a neighbour's cow running through the trial plot. Despite this, the project researchers still feel encouraged by the useful results from the remaining 40 farms.

Another example of the problems which can be encountered in on-farm research is described by a project worker in Kenya: "Farmer Muguga enthusiastically planted the *Leucaena leucocephala* on his land as required for the trial. But he used the land of one of his two wives without consulting her and this proved to be a mistake. This wife didn't like the *Leucaena*, which she saw as a waste of her land, and she complained bitterly, asking why he was spoiling her land and not that of the second wife. The husband, who wanted to avoid serious domestic problems, resigned from the project and destroyed the trial plot."

Some researchers believe that the results of on-farm research will always be contaminated by such variables. Others point out the inevitable disadvantages of on-station research. At this stage, it is probably true to say that most projects would benefit from a well-judged mix of the two approaches.

TECHNICAL PACKAGES

The technical package is the set of innovations proposed by the project for use by local people. Ideally, this should be based on the information on local needs and conditions obtained during the initial surveys and research. In practice, technical packages are often chosen before the project begins.

Some of the characteristics of the main packages commonly used in agroforestry projects are described in the following sections. All have their limitations and some have been badly served by the exaggerated publicity which has surrounded them.

Dispersed intercropping and alley cropping

The principle of intercropping is frequently seen as being at the very heart of agroforestry. Traditional forms of dispersed intercropping are practised by farmers in many parts of Africa. When land is cleared for agriculture, for example, some trees are often left to grow among the crops. Elsewhere, farmers allow self-sown seedlings of particular favoured species to grow within their cropland. *Acacia albida* is probably the best known traditional intercropping species; a wide range of others are also grown, using a variety of different management methods.

Most projects in this survey have promoted intercropping in one form or another. In moist areas, projects have successfully promoted a range of exotic and indigenous species such as *Calliandra calothyrsus*, *Leucaena leucocephala*, and *Tephrosia*, *Sesbania* and *Markhamia* species. In dry areas, *Acacia albida* has been of the species most often used.

Alley cropping is one of the more extreme forms of intercropping. It was developed in the early 1980s, at the International Institute for Tropical Agriculture (IITA) in Nigeria, and involves planting rows (or alleys) of leguminous trees among foodcrops. The trees are pruned during the growing season so that the nitrogen and organic matter in their leaves can be worked into the soil as a fertiliser.

The best results were obtained with *Leucaena leucocephala* and maize, producing very high yields of both wood and maize. This system received wide publicity and as a result has been tried in virtually every imaginable area from hot arid near-desert conditions to cool mountain zones.

An alley cropping component has been included in many of the projects reviewed in this study. In the strict sense, as developed by IITA, it has not proved popular with farmers. One of the problems is that *Leucaena* is not the "miracle tree" which many had considered it to be. Indeed, the view that *Lucaena* will grow anywhere has been so widely accepted that some project staff felt it was their fault that the tree would not grow properly in their area, rather than the limitations of the species.

Another problem associated with alley cropping is the heavy demands it makes on labour and management. A density of many thousands of trees per hectare is required for alley cropping in the form developed by IITA. The investment, the risks and the labour which this requires tend to make the system unattractive to farmers.

But in several projects it was noted that farmers were prepared to take the idea of alley cropping and adopt parts of it that they found useful. In KWDP in Kenya, for example, project staff found that farmers were planting rows of leguminous exotics next to the existing field boundaries. Others planted a

194

single row through their field or on a contour bund. Farmers in the PAP project in Rwanda also planted leguminous trees on contour bunds. In other cases, alleys were established under the guidance of extension workers and evolved into dispersed intercropping over the years.

In several projects it was also found that farmers were prepared to intercrop with leguminous species to provide fodder. The leaves are not returned to the soil but are used to feed livestock, especially by farmers who have adopted stall feeding.

Whether such variations can still be labelled alley cropping is a matter of interpretation. What can be said is that alley cropping in its strict sense was not successful among the projects visited. More dispersed forms of intercropping, on the other hand, seemed considerably more attractive to farmers.

Tree planting on contour lines and farm boundaries

A number of projects in upland areas have developed a package in which trees are planted on bunds constructed along contour lines. The idea is that the trees will help in stabilising the bunds, while at the same time producing wood, fertilising the soil and improving the micro-climate.

Lars Johansson/Panos Pictures

Harvesting fodder from contour strips in the SECAP project in Tanzania; **Grevillea** *trees are planted parallel to the contour bunds.*

In the 20-year-old PAP project in Rwanda, tall timber trees such as *Grevillea robusta* are planted parallel to the bunds; leguminous species are also grown as hedges along the bunds and perpendicular to them. The hedges are cut after 12 months in the first rotation, and after 4-6 months in subsequent rotations. The package promoted by SECAP project in Tanzania is broadly similar.

Experience is showing that some aspects of these packages are more appealing than others. *Grevillea robusta*, for example, has proved very popular and is planted by many farmers along property and field boundaries and to a certain extent within cropland. In some cases, planting with *Grevillea* has led to a visible change in the farming landscape.

Planting leguminous trees on contour lines, on the other hand, has not been so successful. As in the case of alley cropping, the large amount of labour required appears to be a major drawback. Some project staff, for example those in the PAP project, feel that the intensity of land use in their area has not yet reached the level at which farmers would become willing to adopt these practices.

Tree planting in woodlots

Woodlots have long been popular among farmers, especially in densely populated areas where there is high rainfall. The techniques used by farmers have, however, been different from those of conventional forestry. In some cases, they plant the trees in cropland and continue to cultivate around them until they form a closed woodlot, as in a taungya system.

The spacing of the trees is often close and may be as little as 0.5x0.5 metres. Harvesting is usually earlier than in conventional forestry. Most woodlots appear to have been established to produce poles and timber rather than fuelwood.

Several projects, such as RAP in Zimbabwe and the village woodlot programmes in Tanzania, Mali and Burkina Faso, have relied upon woodlots as their main extension package. The initial tendency was to treat the woodlot as a small version of a

This Kenyan farmer has planted several hundred eucalyptus trees around his farm; he plans to sell them for building timber.

Paul Kerkhoff/Panos Pictures

forestry plantation, with eucalyptus being planted in straight lines at conventional spacings and surrounded by an elaborate fence.

The results have almost invariably been poor. Few successful village woodlots survive, usually because of lack of interest from the local community. And although individual farmers may be willing to plant woodlots, they prefer to use their own particular design rather than the rigid recommendations provided by projects. As a result, most projects have moved away from the old-style woodlot approach.

Windbreaks

Windbreaks may seem an obvious answer where strong winds are causing crop damage and soil erosion. The Majjia Valley project, which has been in existence since 1975, has provided the major share of windbreak experience in Africa. The results have been heavily researched but a number of key issues remain unresolved.

There is considerable controversy, for example, about their detailed design. One school of thought is that the windbreak should be low and bushy on the

Harvesting a village woodlot in Burkina Faso; one of the few examples of a successful village woodlot.

windward side with tall trees on the leeward side. This, it is argued, will prevent wind flowing under the tree crown and scouring the land immediately behind the windbreak. Others argue that such closed windbreaks can cause excessive turbulence and advocate designs which give a higher degree of permeability.

One of the problems is that measuring the effect of windbreaks on crop yields is difficult. When yields in a protected field are compared with those of a control plot, the control plot has to be located a long way away to ensure that it is not affected by the windbreak. But this can introduce variations in soil conditions, water availability and other factors which themselves can affect yields.

Research in the Majjia Valley seems to indicate that net crop yields, taking into account the loss of the crop land occupied by the windbreaks, have increased. The gains, however, have been relatively modest and quite variable. It appears that the returns from harvesting the windbreaks for wood products, especially poles, are more important than the increased crop production.

Other benefits, such as the reduction in wind erosion or the general improvement in the environment, are even more difficult to quantify. In the long

The PAF project in Burkina Faso has been highly successful in promoting the use of contour bunds; but the benefits for crop production are the main incentive to farmers, not growing trees.

term, however, these may be extremely important. There are also possible negative influences which must be taken into account, such as the effect of large numbers of trees in lowering the water table. This worry has been expressed by a number of researchers in connection with the Majjia Valley project.

A thorough evaluation of the cost and benefits of windbreaks is thus a difficult task. Up to now there is still not enough experience to say where windbreaks should be used, and what form they should take. It is also worth remembering that large numbers of trees growing in a dispersed pattern over the countryside, as is found in the traditional African farming landscape, can be equally effective in reducing average wind speeds and may in some cases be a preferable option.

Microcatchments

Growing trees in arid and semi-arid areas requires special techniques. One of the most effective approaches is the use of microcatchments. This has been particularly well developed in the Turkana project in Kenya where trees have been successfully grown in areas with an average rainfall of only 180mm per year.

A microcatchment is a shaped depression in the ground which channels run-off water towards the tree planted at, or near, its lowest point. The shape and dimensions of the catchment are varied to suit local conditions. Rock or earth bunds may also be used to help channel the water. Bulldozers and graders have been used by some projects to establish microcatchments, but in most cases they are constructed with hand tools.

Microcatchments are not a new invention. In some farming communities, such as the Dogon in Mali, microcatchments have been used for centuries as a way of improving conditions for crop production. In Burkina Faso, where the PAF project has been promoting microcatchments, there has been considerable interest among farmers in adopting the technique. The beneficial effects on crop yields are providing the main incentive, however, rather than the impact on tree planting.

Promoting natural regeneration in pastoral areas

Because of the vast areas involved, and their low productivity, project packages for application in the dry pastoral regions have to be relatively cheap to implement. Activities such as tree planting and establishing microcatchments are so expensive that they can only be used to a very limited extent. The task is to develop measures which are cheap enough to be widely applied, rather than investing heavily in small areas while the rest is being destroyed.

The single most important measure which can be taken in the drylands is to provide protection against over-grazing. Once this is done, trees and grasses will often regenerate quickly and without further intervention.

Developing and introducing an effective grazing control system is primarily a socio-economic and political issue, rather than a technical one. Projects have used a variety of approaches. Some, such as HADO in Tanzania and Guesselbodi in Niger, have relied upon strictly enforced legal restrictions; the

Turkana project in Kenya, on the other hand, is based upon controls devised and implemented mainly by the communities themselves. Wherever controls have been effective, the impact on regeneration has been substantial. The crucial question is whether the controls can be sustained in the longer term.

Controls on wood cutting can also be important in areas where there is a high degree of exploitation, particularly when this is for commercial reasons. The Turkana project has managed to put a ban on charcoal production, and in the Guesselbodi project in Niger, cutting for commercial fuelwood production is well under control. Again, the effectiveness of such controls depends chiefly on socio-economic and political factors.

Directly seeding large areas with a mix of desirable tree species appears at first sight to be an attractive and low-cost alternative to tree planting. Experience has shown, however, that the costs tend to be high and the results disappointing.

It was tried, for example, in the Ferlo project in Senegal where over 1,700 hectares were tilled and directly seeded as part of efforts to reafforest the areas around wells. The exercise turned out to be costly, the tree growth was poor, and the approach has now been abandoned. A variety of direct seeding techniques were also tried in Guesselbodi for several years but again with little success.

An isolated example where direct seeding appears more promising, however, comes from the Turkana project. There, the indigenous species, *Dobera glabra*, which is highly valued for fruit and browse, is slowly dying out because grazing animals are preventing its natural regeneration. The project has found that by simply throwing seeds among acacia bushes, where they are protected from grazing, they are able to germinate and grow well. The potential impact of this has not yet, however, been assessed.

Effective controls are needed to ensure that grazing regulations work; in the Guesselbodi project in Niger, Tuareg guards have been used to police the project area.

The use of mulch is another relatively cheap measure which can be effective in promoting natural regeneration even in areas where there is a hard laterite crust. In the Guesselbodi project in Niger, it was found that improved regeneration could be achieved simply by leaving branches and other organic material on the ground. Breaking the laterite crust in addition to mulching may help, but is not essential. The mulch brings about a strong termite activity which makes the soil porous and increases the water infiltration.

In contrast with attempts to grow trees by direct seeding, sowing grass seed has been successfully used in a number of projects. In Guesselbodi, some areas, which are judged to be beyond salvage for forest regeneration, are being considered for seeding with grass.

Recognition of the potential for natural regeneration does not mean that tree planting should be totally discarded in pastoralist areas. Farmers and herders may, in fact, be interested in planting a certain number of trees, as has been found in arid Turkana. Such tree planting may not make any great contribution to reafforestation, but it can be a significant step in heightening interest in environmental management.

Matching the package to the context

A clear lesson emerging from projects is that few technical packages can be regarded as truly standard. Most have been developed under specific conditions and need to be tested and adapted if they are to be successfully used elsewhere.

The package must also be matched to the social and ecological conditions of the project area. The approach used in a fertile high-potential farming area needs to be very different from that in a marginal dryland area. This may seem obvious, but a number of projects have had to learn this lesson the hard way.

Furthermore, it is important to bear in mind that there is no such thing as a standard farmer. Economic, social, and cultural conditions vary between different areas and within the same community. The fact that one farmer is interested in planting a hedge does not mean that his neighbour will be equally keen to do so; he may prefer, instead, to grow some fruit trees. The fact that windbreaks have been relatively successful in the Majjia Valley does not guarantee that they can necessarily be replicated in similar physical conditions elsewhere.

Nor should the natural caution of farmers be underestimated. The consequences of a crop failure can be devastating for a family. It is therefore unreasonable to expect a farmer to accept comprehensive new packages involving alley cropping, shade trees, contour hedges, fodder plots and so forth all at once. Such radical changes in the farming system are unlikely to be adopted quickly and, perhaps, for good reasons.

EXTENSION TECHNIQUES

Developing effective extension mechanisms has been a crucial area for all projects, and a variety of techniques have been used. Local cultural factors have played a major part in determining which have worked best in given circumstances; projects have also had to work within the constraints imposed by their budgets and institutional frameworks.

Individual extension

Most projects have relied to some extent on individual extension in which farmers are visited on their own farms by extension workers. One of its major advantages is that it facilitates a dialogue in which the extension worker can often learn from the farmers, as well as passing on advice. Some projects see this as an important element in the training of their staff.

The disadvantage of this approach is that it is slow and expensive. Only a tiny proportion of the farming population in the project area can be reached. There is also a tendency to focus on the more progressive and better off farmers rather than on the poorer farmers who are most in need of help.

Some projects try to widen the impact of individual extension by concentrating on selected farmers whom it is hoped will adopt the new techniques and provide an example to others. Meetings and demonstrations are then held on their farms to which neighbours are invited thus allowing the individual approach to evolve into group extension.

Group extension

Extension services can obtain considerable economies of scale if they are able to focus their efforts on groups of people rather than dealing with them one by one. Projects have a variety of ways of accomplishing this.

The Gituza project in Rwanda, for instance, gives bonus points to farmers who turn up for extension meetings. These entitle them to a reduced price for seedlings as well as an attendance certificate. In other projects, free seeds, seedlings or booklets are distributed at meetings; other attractions have included killing a project goat and providing a free meal to participants, or presenting an audio-visual show.

Alternatively, farmers may be put under pressure to attend meetings. Tobacco farmers in the BAT scheme in Kenya, for instance, are required to attend group meetings and grow seedlings in a group nursery. Extension workers can put pressure on farmers who do not turn up by refusing to provide them with the farm inputs necessary for their tobacco crop. In some projects, particularly those in countries with an authoritarian government system, strong political and social pressure may also be exerted on farmers.

Paul Kerkhof/Panos Pictures

Group meetings do not necessarily have to be organised by the project itself. Sometimes it is better to make use of existing meetings, festivals and other local functions which are likely to be well attended. It is then a question of finding a good way of building an extension element in.

Projects aiming to encourage community action in controlling grazing or

Although there has been a move away from community-based schemes, community action is still essential for some types of project activities; collecting rocks for contour bunds in the PAF project in Burkina Faso.

other activities depend heavily on group meetings. One example is the Turkana project in Kenya, which has set up an elaborate scheme in which thousands of people participate in local seminars on communal land management. A first round of these seminars is held at district and divisional level, involving the administration and important leaders. Subsequent seminars are organised at lower levels to try to reach as many of the semi-nomadic population as possible.

One of the problems which is frequently underestimated in group extension work is the diversity that exists within a community. The term "local people" includes elders, women, children, immigrants, farmers, herders and anyone else who may be living in the village or passing through at that particular time. It is vitally important to identify who are the target groups among all these and ensure that the extension message gets to them. Simply calling a village meeting provides no guarantee that these people will attend.

The Lusume project in Zambia, for instance, found that women were reluctant to attend general extension meetings and it had to set up a series of women-only meetings. In Mali, the Koro project has managed to establish an effective extension scheme which reaches local farmers, but misses out the Fulani herders who pass throughout the region; yet livestock management is likely to have an important bearing on the long term success of the project.

Some projects focus their extension efforts on existing church or women's groups or they may set up special project-related groups. The hope is that, once these groups have been informed, the project ideas will automatically diffuse outwards into the community. Experience shows, however, that this does not always happen. In Kenya, KWDP found that such groups may stick to themselves and exclude outsiders from any discussion of the new ideas.

Model farms and demonstration plots

Model farms have been used by a number of projects. The idea is that they display the project package in a working context so that local farmers can see its purpose and impact, and learn how to implement it.

The model farm may be run by a farmer, a school or a community group. Alternatively, it may be managed by project staff with little involvement from local people. Sometimes excursions are organised so that farmers can come and tour the model farm. But often it is assumed that people will learn from a model farm simply because they live in the same neighbourhood, and see it on a regular basis.

The PAP project in Rwanda has probably made the greatest efforts to use the model farm approach in its extension work. About 100 model farms were created in the project area, each featuring the full package of trees, crops, livestock-keeping and soil-conservation techniques. The farmers chosen to run the farms were heavily supported by extension workers. Free materials were provided and, where necessary, work was carried out with labour paid for by the project.

But even with such a major effort, the extension impact was negligible. None of the local farmers, not even the immediate neighbours, adopted the techniques on display. The model farms remained isolated islands in the project area.

The experience in some other projects has been similar. In EPAP in Kenya, for example, one of the project staff, recalling the handing over of the model farm to the local community, remarked: "The final test was handing over a well-established demonstration plot, with water harvesting systems and everything. None of the local people were interested, nobody took it."

In the RAP project in Zimbabwe, model woodlots were established at nursery sites. The techniques demonstrated included soil preparation and the application of minerals, fertilisers and pesticides. The impact on local farmers seems to have been slight and the activity has now been abandoned.

The idea of having model farms is attractive, since the improvements proposed by the project are clearly visible both to local farmers and visitors. But project experience strongly indicates that they are not effective as extension tools. It may be that the atmosphere which is created appears alien to local people or that the techniques which are demonstrated appear too complicated, expensive and risky.

Demonstration plots are less elaborate than model farms and have been used in various forms by the majority of projects. The plot can be no more than a tiny section of a farm or a small area of ground in which a few tree species are planted.

The advantage of a demonstration plot is that it allows small changes and improvements to be demonstrated in a way which does not make farmers feel they have to alter everything they are doing in order to adopt them. Used in such a way, the demonstration plot is an effective and popular means of extension.

Quite often a demonstration plot starts off as a trial plot. Project staff use the plot to test out the performance of new species. If it is satisfactory, the plot may then be adopted for demonstration purposes.

Using schools to spread the project message

Schools are often seen as a convenient focus for extension efforts. They are found almost everywhere. They are already well-organised by their teachers and involve a high proportion of the total population. Their timetables are also predictable. A project officer may spend days organising a meeting of a women's group only to find it cancelled at the last moment because of a funeral. It might therefore seem as if all a project has to do to achieve a major impact is to find a school with an enthusiastic teacher and arrange to include the extension package in the curriculum.

There are, however, certain limitations when using schools for extension purposes. Projects often find that they are working with only one teacher; if the teacher is transferred the work stops. Neither can children be guaranteed to pass the extension message on to their parents. When KWDP in Kenya evaluated the use of schools in their extension system they found there was little communication between parents and children on tree-related subjects.

One of the hand-made drawings produced by the PAP project in Rwanda to illustrate project techniques.

Many projects have also been disappointed to find that seedlings have died during school holidays.

The effectiveness of schools in disseminating project ideas cannot, therefore, be taken for granted. Nevertheless, it is clear that they do have considerable potential for agroforestry extension. They also provide a chance to influence the next generation of farmers while they are still young. The point is that care is needed in selecting those schools which have a genuine interest in the work and ensuring that they receive the necessary advice and support.

Audio-visual aids

Most projects rely on audio-visual aids as part of their extension strategy. Drawings, slides, film and radio have been used in different ways and with varying degrees of success.

One of the most widely employed extension techniques in West Africa is the GRAAP method which uses pictures stuck onto a flannel board. This has been used, for example, by the Bois de Village projects in Mali and Burkina Faso and also by PAP in Rwanda. No electricity or expensive materials are needed. The drawings, which are used to help extension workers stimulate discussions among villagers, can be locally produced to ensure they are relevant. The method, however, seems to be confined entirely to the Francophone countries.

Slides require a projector, electricity, a screen and a dark space. All this necessitates considerable organisation from the extension staff. Nevertheless, a number of projects have successfully used slides in their extension work. The Gursum project in Ethiopia, for example, has good experience with a simple projector using batteries and a solar-powered recharging system.

The main difficulties experienced in this case have been with the slides rather than the projector. The slides were obtained from abroad and proved to be of little relevance to local conditions. In one unfortunate case, the slide portrayed a photograph taken in Asia of luxuriantly growing *Leucaena*, whereas project trials had succeeded only in producing stunted dwarf trees. No matter what the technique used, it is essential that the extension message is locally relevant.

Drama is an appealing extension tool but it also requires a high degree of organisation. Moving around a troupe of actors and organising the local stage is a big and time-consuming task. Problems also crop up when members of the cast are sick or absent for other reasons.

KWDP in Kenya has used drama for some years and the shows attract large audiences. It has used the technique to raise socially sensitive issues such as the relationship between men and women in families faced with firewood shortages. The project has, however, now given up the use of drama almost completely and replaced it to a certain extent with film.

The films are locally made and performances also attract large audiences. But the demands on organisation and finance remain high, and are out of reach for projects without a high level of external funding. It is also not yet clear to what extent the films have changed tree growing practices or whether they are seen by local people as mainly a form of amusement.

Excursions

Almost all projects use excursions for staff training, but some have also organised them for farmers. PAF in Burkina Faso and HADO in Tanzania, for example, have laid on trips for local farmers and believe they have had a useful impact on the participants.

Excursions are usually popular with extension workers and project staff, which can help in increasing their motivation. This is particularly relevant in the case of government service workers who are often extremely poorly paid. Excursions can also provide an incentive for farmers to participate in the project.

The impressions gained in excursions also tend to be vivid and can influence attitudes greatly. In 1985, the Koro Project in Mali organised an excursion for staff, administrators and farmers to the Majjia Valley project in Niger. The hundreds of kilometres of neem windbreaks made such an impression on the participants that, once back in Mali, windbreaks were seen as the only solution.

For several years after the excursion, the project struggled without success to replicate the Majjia windbreaks. While the project manager had doubts about the windbreak model, he was apparently overruled by the determination of other excursion participants. It was only 3-4 years later that the project began to look for more locally appropriate solutions.

This example shows the powerful impact which an excursion can have. It also shows the need to ensure that the projects being visited are locally relevant. If they are inappropriate or out of context, or if the issues involved are beyond the grasp of the participants, the exercise can prove to be counter-productive.

Involving agriculture ministries

With government-sponsored programmes, it is usually the forestry department that is given the responsibility for promoting tree growing activities. One of the weaknesses of this is that most forestry departments have no tradition of extension work among farmers; neither do they have the large organisations required for mass campaigns to promote tree growing.

In some cases, substantial investments have been made in providing the office blocks, vehicles, equipment and other items needed to build up the forestry department infrastructure. But where agroforestry programmes are going to be mainly concerned with intercropping and livestock management, it is worth questioning whether the forest department is in fact the right institution to be built up.

In a number of countries, there is a growing move to transfer forestry extension work to agriculture ministries. The idea is that agricultural extension staff will then promote trees as simply another crop that farmers grow. Foresters can be brought in as specialist advisors on trees just as there are specialists on coffee growing, soil conservation and other matters.

The idea is appealing. It provides an opportunity to establish an effective extension service without the cost and difficulty of setting it up from scratch. It also gets around the problems which arise when foresters have to play a dual role as forest police on one hand and sympathetic extension workers on the other. There are, however, a number of potential pitfalls which need to be noted.

One of the most basic problems is that many agricultural extension workers have little sympathy with tree growing. In some places, tree growing is actually seen as incompatible with modern farming practices. In Zambia, for example, awards are given to "progressive farmers" who are expected to have cleared all trees from their cropland. In cases such as this, both retraining of extension staff and changes in agricultural policies will be needed if tree growing is to be effectively promoted.

Another problem is that agricultural extension workers are used to working with well-defined and proven technical packages. These simply do not exist in the case of agroforestry. The RAP project in Zimbabwe, for example, attached an agroforestry specialist to the Ministry of Agriculture to train agricultural staff. The impact was slight because of the absence of a clear well-proven agroforestry message.

Despite these problems, the general trend to involve more agricultural and livestock staff in agroforestry is likely to continue. Increased cooperation between foresters and agricultural staff is clearly sensible since the whole essence of agroforestry is to bring about a closer integration between tree growing, agriculture and livestock management.

STAFF TRAINING

In the majority of projects funded by bilateral and multilateral assistance programmes, the project staff have been regular government employees. NGO

Jeffrey Fox/CARE

Centralised nurseries have many advantages, but seedling distribution is often a serious problem; the main nursery of the Majjia Valley Project.

projects, on the other hand, have generally recruited a team specially for the project. Either way, staff have had to learn new skills and techniques.

Recruits have had to be trained, for example, in the implementation of surveys, the establishment and maintenance of nurseries, tree growing techniques, and communication skills. The time devoted to staff training varies from a few weeks in some projects to as long as six months in others.

Many projects use on-the-job training, which may consist of an occasional week-long course, or a series of evening classes spread over several years. Projects such as PAP in Rwanda and the Gursum project in Ethiopia have a strong training component and provide staff with an average of three weeks of training each year. It is interesting to note that most projects which already make a substantial investment in staff training feel they should be doing more.

Manuals are usually seen as an important tool in staff training but their usefulness in practice has varied. Some contain too much technical information which puts off project staff from using them. In one project it was noted that "our manual is more appreciated by visitors than by project staff". It helps if manuals are tested on their target readers before they are published; they also need to be evaluated and revised from time to time to reflect the changing ideas within the project.

Appropriate staff training can go a long way in creating an effective extension team. But there are limits to how much staff can be expected to learn. Some project teams have consisted entirely of foresters who have attempted to deal with matters ranging from livestock management to crop production. This is not realistic. Recruiting a multi-disciplinary team from the outset is almost certainly a more effective approach.

There is also a need to ensure that the curriculum, in both forestry and agricultural training colleges, reflects the need for collaboration between foresters and agricultural workers in agroforestry.

PRODUCTION AND DISTRIBUTION OF SEEDLINGS

Seedling production is a key element in most agroforestry projects. The question of how this is best organised has been a subject of much debate.

Centralised nurseries

Most of the projects in this study have started by establishing centralised nurseries. Some, especially those in very dry areas, have relied on them entirely. The advantage of centralised nurseries is that they enable a seedlings production system to be set up quickly. They also allow the project to control the quality and number of seedlings produced.

The obvious problem with centralised nurseries is seedling distribution. It is often forgotten that, even if farmers consider seedlings useful, they rarely rank them high on their list of priorities. One project found that farmers were unwilling to walk more than about 2 kilometres to collect seedlings. In many projects, however, the average distance between the central nursery and farmers is 20 or more kilometres, and sometimes a lot further. It is therefore not surprising that the uptake of seedlings has often been disappointing.

The problem can be reduced by transporting seedlings to villages, schools and other collection points. But this is only possible if vehicles are available at planting time. Even then, there can be problems for lorries and pick-up trucks attempting to use bad rural roads during the rainy season. As a result of these difficulties, large numbers of seedlings are often wasted. It is not uncommon to find that less than half the seedlings grown in a central nursery find their way to the local community.

Lars Johansson/Panos Pictures

Group nurseries

A number of projects have promoted communal or group nurseries at the village level as a way of decentralising seedling production. The techniques used tend to be simpler than in standard forestry nurseries and they are usually much smaller. Most are supported by projects or government agencies and are provided with plastic bags, tools, seed and advice.

Group nurseries can go a long way towards solving the seedling distribution problem. But projects have found that establishing and sustaining them is not always an easy task. It requires a strong extension organisation and, above all, sufficient interest in tree growing among the local people.

Several projects are now encouraging farmers to raise seedlings on their own farms; this backyard nursery has been protected so that the chickens won't eat the seelings.

Some projects have offered incentives for villagers to establish nurseries. One project provided inputs such as water points or donkey carts and paid for some of the nursery staff. This confuses the issue to some extent as it raises the question of whether people really want the nursery or are just out to get the benefits provided. Projects therefore face a delicate choice. If they provide too little assistance, groups may not be interested in setting up nurseries; but if they provide too much, the motivation of groups may be distorted so the whole operation becomes unsustainable.

Problems of this kind were common among the projects visited. None were entirely satisfied with the results they had achieved so far. In most projects, group nurseries accounted for well under half of total seedling production, and usually much less. Despite this, there was a widespread feeling that group nurseries do play an important role in encouraging communities to learn about tree growing and to take responsibility from seedling production onwards. They can also encourage and promote a group spirit which is extremely important in an overall development context.

Farm nurseries

It has generally been assumed that farmers need to be provided with seedlings if agroforestry programmes are to have an impact. A survey in west Kenya, however, carried out in 1984 by KWDP, found that as many as a third of households were already raising their own seedlings year after year without any outside assistance.

The costs and work involved are minimal. The seedlings are mostly raised at the beginning of the rainy season so that watering is not required. Plastic bags are not used, there is no root pruning of seedlings and little effort is made to protect the nursery against animals. The seedlings are planted on farmers' own lands or are sold or given to others.

Since then, there has been increasing interest in promoting or assisting such 'farm nurseries'. A number of projects now provide farmers with assistance in the form of seeds, plastic bags, technical advice, and in some cases a watering can.

In dry areas, farm nurseries are much more difficult to establish. Their need for watering means they require much more labour to maintain. The fact that tree growing is less attractive also reduces farmers' interest in setting up a farm nursery. The Koro project in Mali has, however, overcome such obstacles and shown good results in its first two years of promoting farm nurseries.

The KWDP project in Kenya has made considerable efforts to promote such farm nurseries. It has used films, radio and mass meetings as ways of reaching large numbers of people. It also provides farmers with seeds. The project has suggested a number of ways in which the nurseries might be improved, but it seems that farmers are quite selective in the advice they are prepared to accept. They seem reluctant to carry out root pruning, for example, despite the improvement it brings in the survival of seedlings after they have been planted out.

of the project. It can also provide a more objective view of where it is succeeding or failing.

In most projects there is a strong case for keeping the design of monitoring and evaluation surveys simple enough so that data processing can be done at the field level without the use of computers. This ensures that results can become immediately available to field staff, short-circuiting the long delays which occur when data has to be sent away to be processed. Alternatively, a two-level system might be used in which a first level of analysis is made in the field and the more detailed analysis is done by computer later.

The most important aspect of monitoring and evaluation is that it should be used to create a climate of open-mindedness and self-criticism within the project. This is not easy, but if it can be done it will mean that the project will be far better tuned to the needs of local people.

Finally, an important, but little considered, question is that of maintaining continuity in the information flow from one set of project personnel to their successors. In some of the older projects visited, there was virtually no information available on previous phases. The lack of such an 'institutional memory' can lead to needless duplication of work or even the repetition of old mistakes.

MEASURING SUCCESS AND FAILURE

Many projects measure their achievements in terms of numbers of seedlings distributed, the areas covered, or how many farmers are visited. For outsiders, this is often the only information by which the project is judged.

While it is obviously necessary to have some objective measure of project achievements, information on physical accomplishments alone often says little about whether the project has brought any worthwhile long term benefits to the people living in the area. This is illustrated in the project to reafforest the areas around wells in Ferlo, Senegal. Thousands of hectares of *Acacia senegal* have been planted, but the impact on local people is recognised as having been minimal.

Projects often have intangible results which are difficult to pin down. In Longa village in Burkina Faso, for example, the villagers took a decision to stop free grazing, so that now all livestock are kept in stalls. This will undoubtedly have a much greater impact than the tree planting. Is this a result of the PAF project and the increased awareness it has created? The question cannot be answered, although an important change is undoubtedly taking place in the village.

Measuring precisely the success of a project is thus difficult if not impossible. But decisions still have to be made about whether new projects should be launched, whether those in progress should be continued, changed, or terminated; and levels of funding have to be decided. Economic cost-benefit analyses can go some way to help but ultimately these are matters of informed judgement.

Lars Johannson/Panos Pictures

The challenge is to create changes that will carry on after project funds have been withdrawn; homemade planting pots in Tanzania made from banana fibre, bamboo, a tin can, and a clay ball.

The Gituza project in Rwanda made a major effort to obtain detailed information not just on seedling survival but also on tree performance. Precise lists were kept of the numbers of seedlings issued and where they went. Extension workers visited the farmers and recorded the survival and growth of the seedlings as well as the location where they had been planted. The farmers were interviewed about their views on the trees and were asked about the species and number of trees they required for the following season. The information obtained was processed by computer and the results were used in planning the future directions of the project.

Such comprehensive and carefully carried out surveys can produce large amounts of useful information. The problem is that it can only be done when the necessary finance, trained staff and computer equipment are available. The system is therefore unsustainable once the flow of foreign funds into the project comes to an end. The same can be said of a number of other projects which have established elaborate computer-based monitoring and evaluation systems.

It is also important to have an open mind and a willingness to look beyond seedling production and survival rates. The Turkana project in Kenya, for example, had to assess changes in the attitudes of local herders towards management of the natural bush as well as finding out how they felt about the Forest Department. One of the key questions in the Majjia Valley project was the effect the windbreaks had on agricultural production. In the HADO project in Tanzania, a major impact of the project has been in areas like nutrition and health.

Monitoring and evaluation is usually most effective when it is carried out by the extension staff themselves. This enables them to see and assess the results of their efforts at first hand. Where monitoring and evaluation is treated as an entirely separate exercise, and is done by outside researchers, extension staff tend to take little interest in the results and the opportunity for useful feedback is lost.

When monitoring is done by project staff, an input from outsiders can be useful when it comes to evaluating the results. This helps to avoid the biases, prejudices and blind spots of those closely involved in the day-to-day running

211

of the project. It can also provide a more objective view of where it is succeeding or failing.

In most projects there is a strong case for keeping the design of monitoring and evaluation surveys simple enough so that data processing can be done at the field level without the use of computers. This ensures that results can become immediately available to field staff, short-circuiting the long delays which occur when data has to be sent away to be processed. Alternatively, a two-level system might be used in which a first level of analysis is made in the field and the more detailed analysis is done by computer later.

The most important aspect of monitoring and evaluation is that it should be used to create a climate of open-mindedness and self-criticism within the project. This is not easy, but if it can be done it will mean that the project will be far better tuned to the needs of local people.

Finally, an important, but little considered, question is that of maintaining continuity in the information flow from one set of project personnel to their successors. In some of the older projects visited, there was virtually no information available on previous phases. The lack of such an 'institutional memory' can lead to needless duplication of work or even the repetition of old mistakes.

MEASURING SUCCESS AND FAILURE

Many projects measure their achievements in terms of numbers of seedlings distributed, the areas covered, or how many farmers are visited. For outsiders, this is often the only information by which the project is judged.

While it is obviously necessary to have some objective measure of project achievements, information on physical accomplishments alone often says little about whether the project has brought any worthwhile long term benefits to the people living in the area. This is illustrated in the project to reafforest the areas around wells in Ferlo, Senegal. Thousands of hectares of *Acacia senegal* have been planted, but the impact on local people is recognised as having been minimal.

Projects often have intangible results which are difficult to pin down. In Louga village in Burkina Faso, for example, the villagers took a decision to stop free grazing, so that now all livestock are kept in stalls. This will undoubtedly have a much greater impact than the tree planting. Is this a result of the PAF project and the increased awareness it has created? The question cannot be answered, although an important change is undoubtedly taking place in the village.

Measuring precisely the success of a project is thus difficult if not impossible. But decisions still have to be made about whether new projects should be launched, whether those in progress should be continued, changed, or terminated; and levels of funding have to be decided. Economic cost-benefit analyses can go some way to help but ultimately these are matters of informed judgement.

One important criterion is whether the project is sustainable. Some projects have clearly had a major impact during their lifetime, but have little prospect of leaving much behind them once they have finished. For those which were designed as emergency relief schemes, this short-term focus is perhaps justifiable. But for projects concerned with long term agricultural development, questions of sustainability are fundamental in judging the success of the project.

A second key criterion is replicability. The problems being addressed by agroforestry are widespread. They will not be solved by projects which cover a few tens or even thousands of hectares. If agroforestry is to have an impact, the role of projects must be to develop techniques and approaches which are spontaneously replicated by local people on a wider scale.

It is important to remember, however, that an element of flexibility is needed even when applying these basic criteria. The economies of some countries, particularly those in the Sahel region, are so dependent on foreign aid that the idea of sustainability can only be seen in relative rather than absolute terms.

A particular example is the Gursum project in Ethiopia, which is heavily dependent on food-for-work. The project activities taking place there may not be sustainable without outside support. But in a country where food aid is fundamental to the rural economy, a project which can develop a sensible food-for-work scheme and, at the same time, promote further development activity which does not depend on food aid, is clearly moving in the right direction.

213

INDEX